Negotiations with Asymmetrical Distribution of Power

AOL Deutschland GmbH & Co. KG
Rechtsabteilung
Beim Strohhause 25
D-20097 Hamburg

Contributions to Economics

www.springer.com/series/1262

Further volumes of this series can be found at our homepage.

Jan B. Kuné
On Global Aging
2003. ISBN 3-7908-0030-9

Sugata Marjit, Rajat Acharyya
International Trade, Wage Inequality and the Developing Economy
2003. ISBN 3-7908-0031-7

Francesco C. Billari/Alexia Prskawetz (Eds.)
Agent-Based Computational Demography
2003. ISBN 3-7908-1550-0

Georg Bol/Gholamreza Nakhaeizadeh/ Svetlozar T. Rachev/Thomas Ridder/ Karl-Heinz Vollmer (Eds.)
Credit Risk
2003. ISBN 3-7908-0054-6

Christian Müller
Money Demand in Europe
2003. ISBN 3-7908-0064-3

Cristina Nardi Spiller
The Dynamics of the Price Structure and the Business Cycle
2003. ISBN 3-7908-0063-5

Michael Bräuninger
Public Debt and Endogenous Growth
2003. ISBN 3-7908-0056-1

Brigitte Preissl/Laura Solimene
The Dynamics of Clusters and Innovation
2003. ISBN 3-7908-0077-5

Markus Gangl
Unemployment Dynamics in the United States and West Germany
2003. ISBN 3-7908-1533-0

Pablo Coto-Millán (Ed.)
Essays on Microeconomics and Industrial Organisation, 2nd Edition
2004. ISBN 3-7908-0104-6

Wendelin Schnedler
The Value of Signals in Hidden Action Models
2004. ISBN 3-7908-0173-9

Carsten Schröder
Variable Income Equivalence Scales
2004. ISBN 3-7908-0183-6

Wilhelm J. Meester
Locational Preferences of Entrepreneurs
2004. ISBN 3-7908-0178-X

Russel Cooper/Gary Madden (Eds.)
Frontiers of Broadband, Electronic and Mobile Commerce
2004. ISBN 3-7908-0087-1.5

Sardar M. N. Islam
Empirical Finance
2004. ISBN 3-7908-1551-9

Jan-Egbert Sturm/Timo Wollmershäuser (Eds.)
Ifo Survey Data in Business Cycle and Monetary Policy Analysis
2005. ISBN 3-7908-0174-7

Bernard Michael Gilroy/Thomas Gries/ Willem A. Naudé (Eds.)
Multinational Enterprises, Foreign Direct Investment and Growth in Africa
2005. ISBN 3-7908-0276-X

Günter S. Heiduk/Kar-yiu Wong (Eds.)
WTO and World Trade
2005. ISBN 3-7908-1579-9

Emilio Colombo/Luca Stanca
Financial Market Imperfections and Corporate Decisions
2006. ISBN 3-7908-1581-0

Birgit Mattil
Pension Systems
2006. ISBN 3-7908-1675-1.5

Francesco C. Billari et al. (Eds.)
Agent-Based Computational Modelling
2006. ISBN 3-7908-1640-X

Kerstin Press
A Life Cycle for Clusters?
2006. ISBN 3-7908-1710-4

Russel Cooper et al. (Eds.)
The Economics of Online Markets and ICT Networks
2006. ISBN 3-7908-1706-6

Renato Giannetti/Michelangelo Vasta (Eds.)
Evolution of Italian Enterprises in the 20th Century
2006. ISBN 3-7908-1711-2

Ralph Setzer
The Politics of Exchange Rates in Developing Countries
2006. ISBN 3-7908-1715-5

Dora Borbély
Trade Specialization in the Enlarged European Union
2006. ISBN 3-7908-1704-X

Iris A. Hauswirth
Effective and Efficient Organisations?
2006. ISBN 3-7908-1730-9

Marco Neuhaus
The Impact of FDI on Economic Growth
2006. ISBN 3-7908-1734-1

Nicola Jentzsch
The Economics and Regulation of Financial Privacy
2006. ISBN 3-7908-1737-6

Klaus Winkler

Negotiations with Asymmetrical Distribution of Power

Conclusions from Dispute Resolution in Network Industries

With 7 Figures

Physica-Verlag
A Springer Company

Series Editors
Werner A. Müller
Martina Bihn

Author
Dr. Klaus Winkler
Blumenstraße 10
60318 Frankfurt/M., Germany
winklaus@gmx.de

This book was submitted as a doctoral dissertation to the Department of Economics of the Friedrich-Schiller-University Jena, Germany, in 2006.

ISBN 10   3-7908-1743-0 Physica-Verlag Heidelberg New York
ISBN 13   978-3-7908-1743-0 Physica-Verlag Heidelberg New York

This work is subject to copyright. All rights are reserved, whether the whole or part of the material is concerned, specifically the rights of translation, reprinting, reuse of illustrations, recitation, broadcasting, reproduction on microfilm or in any other way, and storage in data banks. Duplication of this publication or parts thereof is permitted only under the provisions of the German Copyright Law of September 9, 1965, in its current version, and permission for use must always be obtained from Physica-Verlag. Violations are liable for prosecution under the German Copyright Law.

Physica-Verlag is a part of Springer Science+Business Media

springer.com

© Physica-Verlag Heidelberg 2006
Printed in Germany

The use of general descriptive names, registered names, trademarks, etc. in this publication does not imply, even in the absence of a specific statement, that such names are exempt from the relevant protective laws and regulations and therefore free for general use.

Typesetting: Camera ready by the author
Cover: Erich Kirchner, Heidelberg
Production: LE-T$_E$X, Jelonek, Schmidt & Vöckler GbR, Leipzig

SPIN 11797999   Printed on acid-free paper – 134/3100 – 5 4 3 2 1 0

Foreword

In a globalised world with an ever increasing division of labour, negotiations are becoming an enormously important and complex tool. As they take place everyday and everywhere, it is indeed necessary to know more about their complexity and their associated problems. This necessity is also proven by the growing number of studies related to negotiations and conflicts. With *Negotiations with Asymmetrical Distribution of Power - Conclusions from Dispute Resolution in Network Industries*, Klaus Winkler is providing a vital contribution to this important strand of the literature.

He offers an explanation of why negotiations successfully take in place in situations with asymmetrical access to power and thereby demonstrates that power is a relative rather than an absolute concept. Power differences may well be only perceived, as power is disappearing and reappearing during the process of negotiations. Different cost categories, different layers of negotiations and different strategies lead to the creation of very complicated and non-transparent situations. So, power may be rather the result than the basis of negotiations.

By showing this complexity in a straightforward political economic reasoning and by demonstrating its high policy relevance in a number of case studies taken from network industries, the author demonstrates that asymmetrical distribution of power in negotiations needs no longer to be seen as a problem – but rather as an opportunity. This opens a window of opportunity for negotiating, even in situations, which are perceived to be difficult to handle. This book makes it possible to grasp power in negotiations.

However, the book does not stop there. In the final chapter, Winkler shows that the topic is a political one. Obviously, it is very difficult to lay down rules that allow one to deal with all problems related to negotiation. In particular, network regulation suffers from the dynamics of the market. In other words, the regulator is always a few steps behind the latest developments in the markets. As a remedy, the author discusses a toolbox of support mechanisms which can enable one to run power-negotiations efficiently.

The book benefits from Winkler's exceptional understanding of economic policy. Although it meets high scientific standards, it is also recommended reading for practitioners and government representatives as it offers new insights into the theory of regulation – not only with a view of network markets. This study has been accepted as a doctoral thesis at the Friedrich Schiller-University Jena in February 2006.

Jena, May 2006

Andreas Freytag
Professor of Economics

Contents

Foreword V

1 Asymmetrical distribution of power makes negotiations difficult ... 1

1.1 Efficiency of negotiations in network markets 2

1.2 Asymmetrical power a core issue of negotiations................. 3

1.3 Grasping the complexity of power in negotiations 5

2 Theory of negotiations and power display common characteristics 7

2.1 Negotiations in asymmetrical power markets 7
    2.1.1 The relation between trade, bargain, negotiation and conflict situations.. 8
    2.1.2 From perceived, to resulting, to real power................. 10
    2.1.3 Setup of a power concept in three dimensions 14

2.2 Power-Matrix structures key-characteristics of power...................... 16
    2.2.1 Framework reveals the possibility of access to the different levels of a negotiation.. 16
    2.2.2 Cost forces actors to create options in negotiations 35
    2.2.3 Strategy makes transparency a favourable concept................... 51

2.3 Common characteristics define power in negotiations 63

3 Negotiations in network markets 65

3.1 Network characteristics in telecoms and rail 66
3.1.1 Telecommunication – a dynamic market 66
3.1.2 Rail – a sedate market 70
3.1.3 The need for negotiations in network markets 73

3.2 Power in network market negotiations 73
3.2.1 Framework – the different levels of activity in negotiations 74
3.2.2 Cost – Profit from differences in size 89
3.2.3 Strategy – Coopetition bridges the power-gap 112

3.3 Relevance of the findings for negotiations 127

4 Alternative Dispute Resolution enables efficient negotiations 129

4.1 Alternatives to 'classic' negotiations 129
4.1.1 Is power a problem for negotiations in network markets? 130
4.1.2 Merging the Power-Matrix with guidelines on how to run negotiations – The Harvard Negotiation Project 133
4.1.3 Alternative Dispute Resolution mechanisms support negotiations 141

4.2 Three steps to running efficient power-negotiations 157
4.2.1 Step one: Implement the structural approach of the Harvard Negotiation Concept and the Power-Matrix to boost efficiency 158
4.2.2 Step two: Coordinate the instruments by creating a feedback mechanism 162
4.2.3 Step three: Initiate the use of the new tools 166

4.3 Enabling support processes 168

List of abbreviations 173

References 175

1 Asymmetrical distribution of power makes negotiations difficult

> *"In slightly different terms, the economist's task is simply that of repeating in various ways the admonition, 'there exist mutual gains from trade', emphasizing the word 'mutual'..."*
>
> James M. Buchanan[1]

People and companies have an urgent need to make a good deal. Coming to an agreement demands a negotiation of some sort. Chances are high that a negotiated agreement works to the advantage of all actors[2]; otherwise they would not have agreed to make it. Still there are situations where a problem arises that makes it difficult to reach a common solution. In such situations the actors often feel worse off than before the deal, as they have invested time and resources for 'nothing'.

In an economic analysis this waste of resources is called inefficiency. Efficiency as a scale in negotiations will therefore, firstly, be introduced in this chapter, to enable the later analysis of how negotiations could be handled better. Negotiations in markets where one actor heavily depends on another more often than not seem to lead to problems. It will thus, secondly, be asked if an asymmetrical distribution of power is a core issue of negotiations. It soon becomes evident that the everyday task of making a deal is more complex than would have first been expected. To allow one to grasp the complexity of power in negotiations in an economic analysis, thirdly, a structural approach is introduced. Altogether this can help to gain

---

[1] Buchanan, 1962, p. 354.
[2] From the various terms for participants in power relations Dahl, 1957, p. 203, chooses the term actor. Cross, 1969, p. 7, discusses more terms. I agree that power is a relation among people, not subjects. The term 'actor' describes this best, as actors are actively involved in a negotiation, trying to reach a goal, not playing games. For the thesis I have chosen the term actor as the one that best describes the participants in negotiations. When using the term 'actors' in combination with 'both' the number of actors is not necessarily exactly two. To make it easier to handle, the terms 'actors' and 'both' are here to be understood as a representation of several (most often two) actors negotiating. See Hauser, 2002, p. 16, for a similar discussion.

more insight into the complexity of negotiations in general, hence allowing for more efficient future negotiations.

1.1 Efficiency of negotiations in network markets

Negotiations seem to make sense in general or, in other words, negotiations are an efficient tool for actors. Do negotiations always take place or work out to everyone's advantage, though? By trying to avoid 'difficult' negotiation situations, many negotiations that are (potentially) valuable for all actors involved are prevented from even starting. On the other hand, 'difficult' negotiations can take place and still lead to (good) results. But why do negotiations go wrong? The three main issues that cause problems and thus endanger the efficiency of negotiations will be analysed here.[3]

Firstly, the environment in a negotiation situation, creating different levels of access to the problem for the different actors, may forestall their attempts to focus on the relevant issues. The negotiation itself already consists of many different details and the actors only have a limited processing ability. Together this might prevent the actors from identifying and gaining access to the necessary levels of a negotiation to unleash the opportunities available to them. If then the institutional framework does not (actively) support efficient negotiations, for example by providing the actors with the possibility to rely on external third parties for a process support, it is likely that negotiations fail or deliver inefficient results. This demonstrates the relevance of the interaction between the different stakeholders[4] for efficient negotiations.

Secondly, the actors may not be willing to calculate what the advantages of a negotiation for them are on the basis of a cost-benefit-analysis, or regarding the alternatives that they have. It seems as if actors often are not interested in creating value in negotiations, but rather in defending existing positions. Such a form of behaviour is not necessarily irrational, but one where calculations of short-term effects prevail. This demonstrates the importance of being able to create alternative options in negotiations.

---

[3] Efficiency is here measured in terms of Pareto optimality and will be discussed in more detail below. See Buchanan, 1966, p. 32.
[4] The term 'stakeholders' is here used to refer to all the actors, i.e. individuals and institutions, that have an interest of some sort in the negotiation situation; most likely stakeholders are thus actively involved in the negotiation. In contrast to this definition the term 'environment' includes the stakeholders, but also those that do not actively take part in a negotiation situation.

Thirdly, actors often try to make use of instruments that are available to them. They do so in order to exhibit resistance but without knowing exactly what effect such behaviour will have. Instead of creating a common transparency, decisions are made on the basis of non-transparent individual parameters affecting all actors. This highlights the importance of transparency to allow for individual best choices in negotiations.

Since negotiations are as individual as the actors and the situations in which they find themselves, a focus on the difficulties of negotiations in a special market is appropriate. On the one hand both the telecommunications and the rail transport industries have many common economic properties as they are both network markets. With regard to the negotiation style, however, these two industries are in a different stage of development, which has to do with the liberalisation of these markets. This makes it possible to highlight the different problems of a negotiation in the different time phases. The main question posed here is whether or not negotiations in a network market industry are possible, and if they are, how they can be run in an efficient manner. Such a focused approach allows one to make use of these results and draw conclusions for other markets as well.

1.2 Asymmetrical power a core issue of negotiations

An asymmetrical power distribution in a negotiation is not atypical, whereas perfect symmetry is rather uncommon. If power were evenly distributed between the actors, there would be no need to discuss power concepts: it would be a zero-sum weight in negotiations – existent maybe, but of no relevance. An asymmetrical distribution of power adds weight to the status of one actor, which can affect the course of the negotiations between the actors. When power is distributed in an asymmetrical manner to his disadvantage, the actor will certainly ask himself if he even has a chance of negotiating a good result. The fact is that David was able to slay Goliath at the right moment. David was not afraid of Goliath's overwhelming power but instead used this asymmetry to his advantage. This points to the fact that asymmetry is not a stable state, and it is this fact which makes it necessary to analyse power asymmetries with regard to their sources and dynamics. The question is how actors negotiate under the influence of power and which constraints power imposes upon the negotiators. To answer these questions, power and negotiations will here be looked at as a problem of economic theory. This also sets the structure of the study and provides a link to the key terms of power in negotiations.

Who the actors at the negotiating table are is obvious, but even those not at the table influence the negotiation. To better understand how the stakeholders stand in relation to each other the setting, i.e. the (institutional) **framework**, of the negotiation will be analysed. It could be that only those that own certain property rights are in a position to negotiate. To sum it up: the actors are part of a larger environment based on institutional rights causing an asymmetrical distribution of power between the actors and necessitating negotiation if change is sought. (Key term: framework)

An actor negotiates because he wants to be better off. In network markets this is mostly about gaining access to so-called essential facilities. The actor is sure that he can improve his situation, otherwise he would not enter into negotiations. The actors create their own negotiation situation by making the decision as to when to start and how to participate in a negotiation. This decision is based on the relative position of the actors to each other in an effort to achieve their goals with the least possible **cost**. In some situations, though, actors might not be able to participate in a negotiation, even if they would like to, as a minimum prerequisite amount of power is missing. To sum it up: the actors are interested in exchanging products, services or rights in order to improve their post-negotiation position relative to their position before.[5] (Key term: cost)

The balance of power between the actors only describes the status of short term situations. Actors will review their strategies and the option to cooperate or not before and during each round of negotiations, as the power distribution between the actors changes over time. To negotiate, the actors will use power to improve their relative position. Because of the **strategies** used, the power distribution is not stable, but changes constantly. The individual decision of the actors on how to negotiate is most likely based on experiences of the past which are not necessarily the appropriate scale for forthcoming rounds. To sum it up: the actors plan their actions with or against the others based on the power accessible to them to influence the result of the trade. (Key term: strategy)

The available literature on negotiations describes power as one of the things which very often makes negotiations 'difficult'.[6] It is likely that the existence of power sources creates dynamics and asymmetries which are then a cause of problems or at least a challenge in negotiations. Altogether this makes the asymmetrical distribution of power a core issue of negotiations.

---

[5] See Pen, 1952, p. 24, for a discussion that negotiations are only about the price.
[6] See Fisher/ Ury/ Patton, 1991, pp. 97 ff.

1.3 Grasping the complexity of power in negotiations

The theory developed in this study aims to explain the results of negotiations in markets with an asymmetrical distribution of power. Even though there are, obviously, many reasons why negotiations can go wrong, there are still many possibilities for efficient negotiations that leave the actors involved better off than they were before. The main questions asked here are targeted at finding out if efficient negotiations are possible, even where power seems to be unevenly distributed between the actors. The study uses specific examples to demonstrate the theory developed here, but it will not only look at specific national trends. National differences in the way that negotiations take place have diminished over the years.[7] The overall international experience will be used as a benchmark for further discussion.

Following this introductory chapter, in **chapter 2**, (asymmetrical) power is analysed in a positive manner by combining different theories with an analysis of its impact on negotiations. Hypotheses are set up and then tested with the help of case studies in the following chapter. Like an umbrella, power covers the main problems of negotiations, here summarised with the key terms 'framework', 'cost' and 'strategy'. To be able to grasp the sources and dynamics of power in negotiations these three key terms are structured in three dimensions in a so-called 'Power-Matrix'.

In **chapter 3** the theory will be applied to case studies from network markets. As these markets demonstrate unique characteristics, an overview of the telecommunications and the rail transport industries will firstly be given. Network markets operate within a framework of governmental regulation. This framework has a significant impact on the way that negotiations are handled by the companies. Firms that operate within network markets use and depend on essential basic infrastructures (e.g. a railroad line between two points) which are, by their very nature, scarce and unique. Scarce and unique basic infrastructures can easily provide their owners with a cost structure that may not be available to other players in the industry. Basic infrastructure can easily become a significant cost burden for newer players in a network market. As access to basic infrastructure creates special inter-dependencies between the actors, the actors might make use of strategy to try to improve their relative and absolute position in the game of negotiation.

Chapter 4 outlines and discusses steps that can be taken to make negotiations more efficient. The so-called Harvard Negotiation Project, which

---

[7] See Fant, 1995, p. 177; Huntington, 1996, notes that the differences have diminished only within certain civilizations, i.e. in the so-called western civilization.

is a 'principled negotiations' approach, will be presented as a basis for an alternative dispute resolution mechanism that can help to counter the known problems of negotiations in markets with an asymmetrical power distribution. Merging the Power-Matrix and the Harvard Negotiation Project could provide a guideline for efficient negotiations. In the normative part of the study there will thus be a discussion, by way of conclusion, of how the results from the case studies, together with the theory developed here, can help to set up a general process support for other cases and markets. If the alternative negotiation approaches were to be used, would power (still) be a problem for negotiations? Finally, there will be a discussion of how such alternative mechanisms can be implemented, coordinated and initiated within the existing framework.

Figure 1 highlights the power-structure which is also the red thread of this study.

	Framework dimension	Cost dimension	Strategy dimension
Chapter 2: **Source and dynamics of power**	Section 2.2.1	Section 2.2.2	Section 2.2.3
Chapter 3: **Network market case studies**	Section 3.2.1	Section 3.2.2	Section 3.2.3
Chapter 4: **Alternatives to 'classic' negotiations**	Section 4.1.2	Section 4.1.2	Section 4.1.2

Figure 1: Overview of the power-structure of the study

2 Theory of negotiations and power display common characteristics

What is the relationship of power and negotiations? How can power be grasped in a concept in order to support the analysis of its effects in negotiations? The goal of chapter 2 of the study is to find a workable definition of power and negotiations.

To achieve this goal, firstly an overview of the terms and definitions will be given to provide the ground-work for the theoretical structure of the study. Power is derived from more or less static sources, but the state of power in negotiations is transient and the environment of negotiations is dynamic. To capture both the sources and the dynamics, secondly, a new concept of power is developed and established as the structural element of the study. The so-called Power-Matrix is an arrangement of the different theoretical approaches designed to act as a support for the case studies in chapter 3 and the normative discussion in chapter 4. Thirdly, the common characteristics of negotiations where power seems to play an important role are discussed to conclude with a new power definition for this study.

2.1 Negotiations in asymmetrical power markets

Negotiations in markets where power is distributed in an asymmetrical way might pose special problems for the actors and for the institutions such as regulators that oversee these markets. In this sense asymmetry could be a reason why negotiations fail and are seen as a hurdle to cooperation.[1] The underlying question is "how can weaker actors negotiate with stronger actors and still get something?"[2]

In this section, firstly, the connection between the different terms that describe (difficult) negotiations will be presented in order to facilitate an understanding of negotiations in network markets. Secondly, various approaches to different power concepts in negotiations will be discussed in

---

[1] See Breidenbach, 1995, p. 83.
[2] Zartman/ Rubin, 2000, p. 3.

order to find out which issues need to be subjected to particular focus in the case studies. To accomplish this task a new approach to explain the different states of power and their respective impact on negotiations will be put forward. Finally, both the source and the dynamics of power in negotiations will be structured in a three dimensional matrix. This characterises the attempt of this study to approach negotiation situations where power is distributed asymmetrically between the actors.

2.1.1 The relation between trade, bargain, negotiation and conflict situations

Trade, negotiations and conflict are the terms that describe the situations actors find themselves confronted with when they are dealing. The relation of these terms to each other hints at a dynamic,[3] rather than a static process in negotiations.[4] For that reason these key terms in relation to a negotiation – and the concept of negotiation itself – will firstly be defined.

Trade situations, i.e. the exchange of products, services and rights, are nothing special, but something that takes place every day. It is the goal of every economic subject to arrange allocation of resources in an optimal way and therefore negotiations often seem to be necessary as a support device. Trade is then the general term for the exchange. **Bargain** is a positive sum game,[5] where there is more than one possible agreement and conflicting preferences exist. Altogether a bargain is regarded here as productive, with a distributive purpose.[6] **Negotiations** 'organise' cooperation and take place in order to make transactions and the allocation-process possible. They are a preliminary condition for the exchange of goods, services and/ or contractual rights. Thus negotiations are a part of the trade, if not the heart of trade. The difference between bargain and negotiation is that a bargain describes the formation of demand, whereas negotiation describes the whole process of trading (i.e. also the necessary communication between the actors), including the bargain.

That there are so many different ways to describe trade clearly displays the everyday character of negotiations.[7] Everyone seems to know what a negotiation is. Many think of bazaars in Middle Eastern countries when hearing the term negotiation, but forget the silent negotiations that take

---

[3] See Ståhl, 1972, p. 13, for a graphical overview of the relation of the different terms.
[4] See Lyon/ Huang, 2002, p. 112.
[5] See Deutsch/ Krauss, 1962, p. 52.
[6] See Cross, 1969, pp. ix and 7.
[7] See Carraro/ Marchiori/ Sgobbi, 2005, p. 48.

place every day in the supermarket around the corner, i.e. whether or not to make a purchase based on the price tag on a product. The cultural differences in trade are one reason why they have been looked at from many different (research-) perspectives.[8] Even though there are numerous and ancient cultural versions of negotiations,[9] covering a diversity of issues and with many actors with different quantities of power, a typical structure and common mechanics of all negotiations can be identified.[10] Fun is not usually the driving force why individuals enter into negotiations.[11] It is rather that they hope to be better off after the negotiations than they were before, despite the expenses involved in a negotiation.

Actors start out from different positions, such as, for example, in a simple case, where the purchasing actor would like to pay a low price whilst the supplying actor would like to realise a high price. Is therefore every trade a situation of **conflict**? Since the term conflict has destructive connotations in the context of negotiations,[12] conflict situations in this study are those where a zero-sum game is expected: if there is a winner there must also be a loser.[13] Negotiations are rather a way of harmonising the different (if indeed they are different) goals of the actors by, for example, achieving a certain output. In both conflict and negotiation situations the actors as rational subjects will make use of all possible powers at hand in an attempt to influence and to benefit from the trade.[14] Thus conflict is not a prerequisite for a negotiation, but could be the starting point. On the other hand, if a conflict arises, problems with the negotiation are most likely to be part of this conflict.[15]

Negotiations are a way of harmonising the different – but not necessarily conflicting – interests of the actors in a trade situation. The term negotiation in this study is used to describe a method of exchanging information between actors with the goal of reaching an agreement that allows both to be better off afterwards than they were before, no matter what the starting point, the cost or the dynamics of the situation. The extremes

---

[8] See Ehlich/ Wagner, 1995, p. 1. They show that there are four main areas of research concerning negotiations: training, ethnographic, psychological and economics. Underlying all four is the communication process.
[9] See Rehbein, 1995, p. 70.
[10] See Schelling, 1956, p. 290 ff.; Wagner, 1995, p. 12.
[11] See Pen, 1952, p. 28.
[12] See Alexander, 1998, p. 38.
[13] See Charles, 1995, p. 151; Schelling, 1956, p. 281.
[14] See Lewicki et al., 1994, p. 8.
[15] See Hauser, 2002, p. 17.

within a negotiation are cooperation and conflict;[16] the likely case of a negotiation is something in between.

2.1.2 From perceived, to resulting, to real power

There is not just one "label for power"[17], a fact borne out by the many different definitions that exist.[18] This makes power a vague concept.[19] Besides economics, psychological and behavioural sciences are probably the disciplines most involved in the studies of power.[20] Most often the term power is used to articulate negative feelings, in such terms as 'abuse of power'. "For some, power permits selfish gain, for others power encourages nobler pursuits."[21] It is thus hard to find a sure-fire indicator as to which use of power is good or bad. Power is hence regarded as one concept, with different sciences to do the analysis. Here it is important to look at power since it is the basis of the economic markets.[22] "By power is generally meant participation in decision-making and the bases thereof, such as property rights, income, position, influence, and other rights of economic significance."[23] In that sense power is participation in the markets. In the present study this kind of participation is narrowed down to the concept of negotiations. Since power is the key term and object of this study it is necessary to examine power in more detail. To be able to 'use' power, some of the different approaches will be presented here in order to provide a structure for the different terms and definitions. The goal is to make it possible to find the best concept for this particular study. It will be necessary to identify those characteristics of power that are part of negotiations.

Hart distinguishes between the following three approaches to describing power:[24] power as control over actors, power as control over resources and power as control over events and outcomes. **Power as control over actors** describes the situation where one actor intends to produce movement by

---

[16] See Zartman, 1974, p. 388.
[17] Dahl, 1957, p. 201.
[18] See March, 1963; Greene, 2000, gives an overview of the term and the use of power over time.
[19] See Rothschild, 1971, p. 15.
[20] See Zartman/ Rubin, 2000, arguing that the different sciences make a definition of power even more complex.
[21] Seeley/ Gardner/ Thompson, 2002, p. 15.
[22] See Warren, 1979, p. iii; Young, 2002, p. 48.
[23] Warren, 1979, p. iii.
[24] This paragraph builds on Hart, 1976, pp. 289 ff.

the other.²⁵ It is the ability of one actor to impose his will against the will of the other.²⁶ While trying to produce such movement, one may depend on the power to bind oneself.²⁷ The shortcomings of this approach are that it is not clear whether or not an investment to establish power is needed first. **Power as control over resources** is the most prominent definition and probably the most visible kind of power. Masses of resources can be seen and displayed: size, number of employees, money and resources available are notable and indicate why one actor is perceived as being more powerful than the other. This definition is very descriptive, but not at all correlated to the result of a negotiation. The shortcomings of this approach are that powerful is likely also to be complex, slow, and inflexible. Thus power as control over resources might make the actor inflexible and consequently weaker on the one hand and yet stronger as he has a large amount of resources at his disposal. The result of these conflicting effects is often unclear. **Power as control over events and outcomes** puts primary focus on the results, instead of on the actors. The actors cause these results, but since they have only partial control over the events taking place, they do not know how much power they have until they see the result.²⁸

There are advantages and disadvantages with the power approaches discussed above. In the first approach the relationship between the actors, a function of their environment, is in focus. In the control over resources approach the spotlight is on the actors' positions, measured in terms of available resources. For the third approach only the result of the action counts. Hart believes that the concept of power as control over events and outcomes is superior to the other approaches, because it can cope with interdependence. Interdependence means that the actors depend on each other for a solution.²⁹ The disadvantage of this approach is that the results can only be measured retrospectively and thus cannot help in advance to define a strategy for a behaviour.³⁰ In addition the necessary definition of a finite set of actors and events seems not to be realistic.

The above theoretical descriptions of power attest the need to find a **workable power definition**. Since it was demonstrated above that the major difficulties of a definition seem to lie in describing power as such (where does one derive power from?) and the change of the actors' power

---

[25] See Zartman/ Rubin, 2000, p. 8.
[26] See Pen, 1952, p. 29; Dahl, 1957, p. 203; Hart, 1976, p. 291.
[27] See Schelling, 1956, p. 282.
[28] See Schelling, 1980, p. 22.
[29] See Lewicki et al., 1994, p. 24.
[30] See Maoz, 1989, p. 240.

positions over time, key elements of the definition of power in this study are the **source** and the **dynamics** of power. This will be reflected in the following power-concept for negotiations, based on 'perceived power', 'resulting power' and 'real power'.

Actors are equipped with a 'starter package' of power **sources**,[31] from which they derive power for their individual negotiation situation. The term market power (often called SMP, the abbreviation for 'significant market power') describes the common idea of power in the economic context: actors have access to a certain amount of financial and human capital which they are able to invest in the negotiations.[32] Implementing power in negotiations gives "the negotiator some *advantage* or *leverage* over the other party."[33] To actors it appears at first as if they could make direct use of this power, but in the end they do not know exactly what they can cause with it.[34] Uncertainty makes it impossible for actors to conclude the exact effects of power ex ante. This kind of power will here be called '**perceived power**'.[35] In one sentence: perceived power is what actors expect to be able to deliver against the other actor in a negotiation (ex ante, i.e. in time-period 0)

Nevertheless power can and will be put to use by the actors in order to try to generate results,[36] even though most of the actors are aware that the use of power at this stage does not lead to certain results at all.[37] The employment of and the possible access to resources might have an effect on the result of the negotiation, but to what extent is unknown. This makes it difficult for actors to understand what kinds of power they have access to. On the one hand the **dynamics** of the use of power in negotiations lets actors, looking at the same situation from different perspectives and with a different feeling for time, expect different effects of their power actions. On the other hand the negotiation will definitely have a result which mirrors '**resulting power**'. In one sentence: resulting power can only be measured after the negotiation (ex post, i.e. in time-period 1).

Adopting either the ex ante or the ex post perspective on power, i.e. a static and one dimensional view,[38] does not help to understand the process of negotiations better. Only when actors make use of their **interdepen-**

---

[31] See Lewicki et al., 1994, p. 298.
[32] See Simon, 1957, p. 71.
[33] Lewicki et al., 1994, p. 292.
[34] See Warren, 1979, p. iv.
[35] See Simon, 1957, p. 71. He calls this 'formal power'.
[36] See Haley, 2002.
[37] See March, 1963.
[38] See Young, 2002, p. 50.

dence do they have access to '**real power**' and even then only for certain issues and only within the timeslot which is delimited by the ability of the actors to pool their powers. Only in this way are they able to internalize the dynamics of power in negotiations and to limit the uncertainty about information and time. By taking learning effects over time into consideration it is possible to grasp power even in a dynamic negotiation environment. In that way power can be implemented by the actors together as an instrument.[39] In one sentence: real power is available only 'issue specific' and for as long as the actors build upon their interdependence (available at any time, i.e. in time-period 0, 1 and also as a basis for following negotiations also in 2).

Figure 2 connects the key terms developed here in one picture.

Reason	Uncertainty about one's own position leads to	Implementation of resources leads to	Instruments applied together lead to
Kind of power	**perceived** power	**resulting** power	**real** power
Timeline	ex ante = 0	ex post = 1	any time = 0, 1 or 2
Effect	One's own expected **source** of power is delimited by that of the other actor.	The **dynamics** of power lead to unexpected actions of the individual actors.	Any time **interdependence** between the actors is realised they can grasp the source and dynamics of power.

Figure 2: From perceived, to resulting, to real power in a negotiation

Source: own design

The difficulty is that power is based on perception, but also produces concrete results. That 'small' actors, i.e. those that do not have much **perceived** power, might win over 'big' actors and hence constitute more **re-**

---

[39] The terms 'instrument' and 'tools' will here be used with regard to the usage of power. I understand these terms to be interchangeable and will use them that way. Both are support gadgets to transport or execute a certain wish of the actor. It is not necessarily the case that the 'translation' works, i.e. the instrument might not exactly do what the actor wanted it to do, but might have side effects as well. In general an actor first defines his needs and then decides on which instruments to choose in order to fulfil his demands best.

sulting power mirrors the paradoxical nature of power definitions.[40] Thus power does not have the desired effect as long as "the power of power"[41] is uncertain. To enable one to make use of **real** power a model considering the sources and the dynamics of power is necessary.

2.1.3 Setup of a power concept in three dimensions

The goal of this study is to find a way to make best use of the resources available (efficiency). It is not so much the question as to whether negotiations should take place at all (question of effectiveness). That is why the way the success or improvement is measured here is in terms of efficiency.

In the following a structure of power in negotiations will be set up in a way which mirrors the approach toward a negotiation in just the same way as an actor would do, bearing in mind the aforementioned comments on power. This theoretical approach will help the research of negotiations in markets where power is distributed asymmetrically. The underlying assumption is that asymmetry makes power in negotiations a relevant issue.

At a later stage this approach could also help the actors in their negotiation situation as a guideline. How can they improve the efficiency of their negotiation? By choosing this perspective the structure of power in negotiations can be described more intuitively. For that reason the characteristics of power and negotiation are developed in the dimensions 'framework', 'cost' and 'strategy', which are the cornerstones of the so-called Power-Matrix. These three dimensions are set up to represent and structure the process of negotiations. It is as if an actor were to discover his limits and options as he negotiates.

In dimension one a **framework** is developed in order to explore the different levels of power. The source of power in the framework dimension is based on laws, values and roles, budgets and behaviour, together with the distribution of property-rights.[42] These might be restrictions or opportunities for the actors and allow for different degrees of access to power in negotiations. The dynamics of power give the actors the opportunity to decide on which of the different levels to act, i.e. on that of the actual negotiation or of the environment.[43] Using the theory of public choice as an

---

[40] In this study the terms 'big' and 'small' (actor) express the power-asymmetry between actors.
[41] March, 1963.
[42] See Voigt, 2002, p. 17.
[43] See Schelling, 1957, p. 20, demonstrating that action does not necessarily mean that an actor has to do something actively. Inaction is just as well a form of action.

indicator, the reason for the limited abilities of actors to act and react is discussed. How does the actor gain access to the different levels of a negotiation?

In dimension two the term **cost** of power is developed. The source of power in the cost dimension is based on the transaction costs of the negotiation. Different access to so-called essential facilities creates 'big' and 'small' actors, where the actors do not necessarily know whether they are actually 'big' or 'small'. The dynamics of power are accelerated when actors try to create more certainty, trying to gain control of the negotiation situation. This acceleration causes uncertainty but could also help the actors to make transparent their interdependence allowing both 'big' and 'small' actors to profit from asymmetries.

In dimension three the term **strategy** is developed. The source of power in the strategy dimension of power is based on the information and the timing of the possibilities available to the actors – and the data which the other actors can only imagine but do not know for sure. The dynamics of power pick up speed as a result of the actors' actions when they try to react or 'pro-act' to the activities of the other actors. The actors try to steer the result of the negotiation by making use of their strategic possibilities. The implementation of strategies can cause shifts[44] of power in the negotiation, with contradictory results, but still the actors make their individual decision to cooperate or not to cooperate with the other actors.

Figure 3 graphically illustrates the three dimensions of power and the link to negotiations – establishing the Power-Matrix as the structural element of the study.

	Framework dimension	**Cost dimension**	**Strategy dimension**
Source of power is based on ...	property rights distribution	essential facilities	time and information
Dynamics of power lead to ...	the need to access different levels of the negotiation	uncertainty and non-transparency	cooperation and non-cooperation

Figure 3: The three dimensions of the Power-Matrix

Source: own design

The Power-Matrix can, obviously, only provide a snapshot of one point in time. Dimension one describes the starting point of an actor. In the

---

[44] With 'shifts' in this study shifts of power in a negotiating situation are described.

course of a negotiation it may well be that the actors 'pass through' this matrix several times. Actors do usually expect several rounds of negotiations, as it is assumed here that negotiations are based on a long term relationship between the actors. The advantage of describing power as a process is that problems within an asymmetrical power distribution in negotiations can be demonstrated in detail. The three dimensions of the matrix will allow one to fully analyse the problems that occur during a negotiation process.

2.2 Power-Matrix structures key-characteristics of power

In this chapter power and in particular the effect of asymmetrically distributed power in negotiations is analysed. The main question is whether the source and/ or the dynamics of power cause problems, which make negotiations inefficient.

The chapter is structured as follows: within the three dimensions of the Power-Matrix, each of the key terms framework, cost and strategy will be reviewed in detail so that negotiation situations where power is distributed asymmetrically can be analysed. The three dimensional model allows one to observe the different dimensions at the same time as it shows what theory can do to explain negotiations.

2.2.1 Framework reveals the possibility of access to the different levels of a negotiation

Where does an actor stand in a negotiation? How can he position himself to make sure that he achieves good results in his negotiation?

The **framework** seems to play an important role in answering this question. For that reason firstly, the so-called property rights will be defined as the starting position of an actor in a negotiation. There are different levels of a negotiation: the actual negotiation and the environment, for example. The way in which these levels coexist and how they are manifested will be examined. Secondly, there will be an analysis of where and how (power-) changes on the different levels of the framework come about. It will be important to find out how actors behave in negotiation situations and what kind of (power-) dynamics they cause. Access to the different levels could make the difference between an actor becoming powerful or not. Finally, and as an initial résumé, the question as to whether a starting point in negotiations is necessary in order to keep track of the changes in the distribution of power during the negotiation process will be discussed. This could possibly help to understand how power is created and what the use of power does to a negotiation situation.

Property rights are a source of power

Negotiations are necessary for trade, but to be able to run negotiations efficiently, a framework of rights is needed.[45] Every actor needs such a framework to be able to make use of power in negotiations, but power as such does not deliver a framework.[46] Power can only work within a given structure which functions as an anchor, where power provides the 'weight' of the anchor. In the following section the frameworks are analysed and the restrictions and opportunities which arise from them will be explored. Power asymmetries between actors could be caused by the framework.

Values and soft factors function as an anchor. The framework includes or even delivers what is here called a 'starter-package' of values and rights; this is the anchor of power in negotiations. Economic models often exclude values, though, as they are soft factors and hard to define.[47] Based on personal judgment and experience such soft factors make the difference for the actor between a good and a bad deal. Legal power might seem unfair to the one that 'loses' and at the same time it might seem fair to the 'winner' of a court case.[48] There is no 'black or white' or 'true or false', but still human beings have some kind of notion of good and bad and it seems necessary to make such a differentiation in the context of analysing power in negotiations. The historical background 'forms' people and consists of religion, structure and education. Everybody has a certain understanding of human rights and dignity and they are entitled to expect that everyone else should respect these sentiments. If somebody intentionally breaks these basic rules one would describe it as an abuse of power. Power, in a certain way, is included in our society and accepted as a means of order. The expression 'group pressure' implies that structure always guides (or rules) people. So there is probably in every system, state or group the willingness to give up some of one's own rights for the sake of the group. In a social or even economic sense this would be called 'for the profit of the group'. One gives away some power to the system or the structure to have a certain amount of order and to ensure that one will be treated in a correct way. At the same time this creates one part of a system which exerts power on everyone within it.

---

[45] See Engel, 2002b.
[46] See March, 1963.
[47] See Rehbein, 1995, p. 69; Simon, 1957, pp. 63 ff.
[48] See Rawls, 1971; Mähler/ Mähler, 2000, p. 11; Aufderheide, 2000, p. 156.

Power is linked with value judgments,[49] as the "important characteristic of any game is how much each side knows about the other's value system."[50] Power makes the values fit together,[51] and values deliver a relative scale for the measurement of power.[52] This is supported by the finding that the longer actors negotiate, the more predictable the results are; they can probably measure the effects of power more realistically after some time.[53] In economic theory the actors know if they have made a good deal only after they have signed the contract. As values can provide a first indicator for the framework dimension, they serve the actors as an anchor for making decisions in negotiations.[54]

Rules of the game connect the different levels of a negotiation. By staying within certain limits actors assume certain roles, for example, monopolists are not allowed to employ some tactics others may.[55] Even though differences between cultures exist, numerous behavioural expectations remain in every culture. There is a difference between common rules, context specific rules, codes of conduct and those issues that are beyond the reach of law but have nonetheless a restricting influence on negotiations.[56] First the effect of rules will be described.

Most actors stick to the established rules. They do so because they are expected to behave that way.[57] This shows that power is already being exerted just because of the existence of a framework. The framework of a negotiation alone might thus be responsible for a certain distribution of power between the actors. In this respect negotiations take place under a given set of rules. The framework of a negotiation works just like an umbrella, keeping everything that is taking place during and around a negotiation under one roof. But still, it is difficult to use theory alone to describe, in a clear-cut way, what must be included in order to delineate the complete setting of a negotiation within a model. Leaving out certain parts of the environment might distort the result, whilst considering too many parts might make the case more complicated than necessary. The framework, furthermore, is not necessarily (fully) visible to everyone, but exists only as a virtual construction. It describes the setting, can change during the ne-

---

[49] See Charles, 1995, pp. 151 ff.
[50] Schelling, 1958, p. 233.
[51] See Zartman, 1978.
[52] See Zartman/ Rubin, 2000, p. 7.
[53] See Druckman, 1971, pp. 503 ff.
[54] See Blount, 2002.
[55] See Shapiro/ Varian, 1999, p. 303.
[56] See Korobkin/ Moffitt/ Welsh, 2004.
[57] See Maoz, 1989, p. 244.

gotiation, and is not necessarily obeyed by the actors at all times. The framework seems to be a mixture of inconspicuous borders that the actors sometimes touch but try not to cross: a complex mixture of written and unwritten rules that are different for each and every individual situation.

Rules for the negotiation provide the 'glue' that makes the different levels (actors that are, obviously, part of the negotiation and those that influence from the 'outside') stick to each other. The amount of glue used describes how well the two levels are coordinated. The glue itself consists of many soft factors that are part of the negotiation, such as politics and lobbying, and which have different levels of stickiness. From this perspective the negotiation can be pictured as a whole consisting of different parts. These different parts are glued together, but not necessarily permanently, and can be rearranged as necessary.

The Schelling framework – How the rules of the game are set. Schelling has fairly 'exotic' ideas about how the rules of the game interact with negotiations and power.[58] He argues that borders (of a framework) require agreement or at least recognition. That implies that rules for a negotiation need to be set, not necessarily by the participants themselves, but certainly they have to be 'posted', so that everyone can take notice of them. But what counts as posting of the rules? Schelling notes that posting can be accomplished without communication, if only 'obvious' decisions are made. Direct communication is not needed as long as the actors have some kind of framework, albeit simply the rules of the game, to stick to. In this case, people can align their intentions with others'.

The rules of the game can be conventional wisdom, which allows the actors to make 'obvious' decisions which are the same everywhere. In bargaining situations this is more often than not a simple matter of an equal share agreement. But of course an 'obvious' outcome depends greatly on how the problem is formulated. "Each party's strategy is guided mainly by what he expects the other to accept or insist on; yet each knows that the other is guided by reciprocal thoughts. The final outcome must be a point from which neither expects the other to retreat."[59] Actors develop a system of expectations,[60] which can lead to a seemingly endless circle, where no actor knows any longer what action causes what. Such a system of expectations is captured in the so-called 'prisoners' dilemma'. In order to converge on a single point, actors meet at 'the middle ground' in negotiations. Only a recognizable limit is needed, as an obvious place to stop. Even

---

[58] This section builds on Schelling, 1957.
[59] Schelling, 1957, p. 29.
[60] See Coddington, 1968, p. 1.

without communication or negotiation these limits can be found, as can be shown in the most extreme case: war.

These findings can be summarised as follows: simple and unambiguous 'answers' to questions, sometimes not even asked, are needed. Transparency in the formulation of the problems and definition of the possible solutions can help to reach this goal. To enhance the likelihood of a successful outcome, a visible framework, if not acknowledged ahead of time, is necessary. Even a poor signal may command recognition in default of any other. External independent third parties might be useful in this respect to point out the rules of the game without being involved in the contents of the matter under discussion. Even untested unilateral suggestions may prove beneficial when there is no alternative. Those actors that are not interested in an efficient solution, i.e. actors that actively want to promote their divergent interests, are motivated by a desire to destroy communications.

The intuitive behaviour of actors and the way the rules are communicated are based on the framework of values and give the actor a role in a negotiation.[61] But it also works the other way around: the moves of the actors reveal information about the players' value system.[62] This means that there are some parts of the framework that seem to deliver the support for the 'whole system'. These basic parts work as a point of reference. Intuition tells us that every actor searches for such a point of reference when making his decisions within a negotiation. Actors need such an anchor to be able to make decisions within their horizon of experience. An anchor can also serve as a starting point. Actors look for something familiar and start from within their own framework to use it as a shield to protect themselves against new and excessive information which cannot be otherwise avoided. Thus the framework will be used by the actors as a scale by which to measure their different decisions and the actions they are confronted with, as a basis for their next steps.

Property rights institutionalise the framework. The rules of the game, together with the 'weak' factors such as external shocks,[63] culture[64] and emotions, make up the framework for negotiations. Even though as single reasons they may be weak, together they have a substantial effect on negotiations.

---

[61] See Schelling, 1958, p. 209.
[62] See Schelling, 1958, p. 215; Druckman, 1971, p. 533.
[63] See Frey, 1991, pp. 11 ff.
[64] See North, 1992, pp. 43 ff.

But how is the framework manifested? Coase has connected the different aspects. In his article "The Problem of Social Cost"[65] Coase describes the paramount importance of considering that one actor might have the 'right' to do certain things, whereas the other does not have that certain legal right.[66] To illustrate how the framework is established, imagine that at first no man-made framework exists except for the natural limits. Over time a framework develops and is expanded by man, often under the direction or orders of someone, but also via negotiations in which the different stakeholders try to realise individual goals. The initial set of rules and laws can only be changed using exactly those rules which are part of the initial set and laws. This makes the property rights distribution between the actors the 'starter package' for negotiations.

The theory of property rights builds upon the knowledge that contracts are incomplete: not every eventuality can be incorporated ex ante and therefore contracts can only cover some of the situations that are possible. In that sense information is put into the centre of the contract theory as trade partners have different levels of (access to) information, depending on the institutional setting and the different property rights. Thus before an exchange can take place the actors need to find a way of defining the contractual rights, the exact dimensions (time, place and quality) that are valid for this one exchange. Depending on the distribution of rights, the result may be more favourable for one actor than for the other; if the rights are distributed differently, the result might be turned around.[67] Hence the distribution of property rights has a major influence on the result of a negotiation and makes it necessary to look more closely at those institutions that set up and define the property rights.

Even though there are different schools of institutionalism,[68] for all of them it is true that institutions in one way or another set a frame and function as a filter for the actors.[69] Institutions take over responsibility from the actors[70] and mostly derive their power base from the law.[71] The difference lies in how these institutions are set up and how they work as a framework for negotiations between the actors. If the state does not provide such a framework this is called 'government failure', as the state does not deliver a basis for the actors to make use of the resources in an efficient way in the

---

[65] Coase, 1960.
[66] See Coase, 1988, pp. 12 ff.
[67] See Coase, 1988, p. 115.
[68] See Eberlein, 2001, p. 356; Furubotn/ Richter, 1997, pp. 31 ff.
[69] See Ghertman/ Quélin, 1995.
[70] See North, 1992, p. 3.
[71] See Mähler/ Mähler, 2000, p. 10; Foucault, 1977.

course of a negotiation.[72] Even if the state supplies an institutional framework the actor still has a choice to make. Coase highlights the fact that each actor's individual decision model needs to include the whole institutional framework of the trade.[73] Using the example of a stock exchange he notes: "It suggests, I think correctly, that for anything approaching perfect competition to exist, an intricate system of rules and regulations would normally be needed."[74] To Coase the theory of trade is really a theory of choice:[75] the participants in the exchange have the choice of different institutional settings which result in different cost structures for the individual, but not the whole trade. The individual actor must make a choice (of existing) institutional frameworks, as it is not enough to assume that the economic system simply works by itself.[76] This means that if a single actor cannot change the framework, he still has the option of choosing the institutional setting that he would prefer, by moving himself (often referred to as 'voting with one's feet').

The so-called 'constitutional contract'[77] sees the co-existence of institutions and individuals as a means of describing and securing the distribution of rights, i.e. establishing a framework. Such a framework works almost like a public good: everyone uses it, but no one wants to pay for it. The cost of the framework is not zero, but it cannot be calculated in a reliable or exact way. The individual must then decide if he is willing to pay the necessary amount for a good that he cannot provide himself, but where there is no market that sets a price either. For organisations this is seen in a somewhat different way: organisations exist to arrange structures and try to establish and change rights so that they can profit from them.[78] At the same time the development of these organisations depends on the institutions.[79] This is no different for network markets: "Institutional economics underline the importance of the contractual dimension to the functioning of network industries, and thus to their reform."[80] The institutional framework, within which the negotiations take place, can be influenced even by the single negotiator. A good actor "must be able to not only follow rules but also to *set them*."[81] The actor has an interest in influencing or creating a

---

[72] See Donges/ Freytag, 2004, pp. 181 ff.
[73] See Coase, 1988, p. 3.
[74] Coase, 1988, p. 9.
[75] See Coase, 1988, p. 2.
[76] See Coase, 1988, p. 34.
[77] See Demsetz, 1988.
[78] See Vaubel/ Willet, 1991, p. 1.
[79] See North, 1992, p. 5.
[80] Glachant, 2002, p. 308.
[81] Young, 2001, p. 17.

framework, so that his goals can be reached as quickly and easily as possible. The amount of change that a single actor can create varies greatly. It partly depends on where, how and how much power is applied to cause change. It is sometimes argued that institutions are shaped by those that have negotiation power.[82] Besides making a choice between different existing frameworks, there are possibilities of influencing the framework, too, as the institutional setting cannot be taken as given or granted, but as something which remains dynamic.

To sum up the connection between property rights, the institutional framework and negotiations it can be stated that not only individual success but also the overall result of a negotiation depends on the way the property rights and the institutional framework are set.[83] In order to profit from the framework it is necessary for both institutions and the individuals to invest in it. All the elements of the market, i.e. the individuals plus the institutional setting, may alter the results of individual trades. This demonstrates that institutions determine the cost of trade. The framework is therefore a source of an asymmetrical power distribution between the actors.

Regulation as a framework mimics the market and creates institutions. Regulation acts as an institution, which tries to imitate the market and eliminate market power at the same time.[84] Regulation is seen as 'competition put on the stage' with the help of rules.[85] It is based on the understanding that a certain market (or a segment thereof) needs support because of problems that exist within the structure of the market. Those problems are called market failures.[86] It is assumed that the market mechanisms themselves have been unable to keep up a functioning market: free competition is not possible. The goal of regulation is to run the markets efficiently. Regulation poses an economic dilemma, though, as it tries to achieve several goals at the same time, i.e. efficiency and social welfare, which may at times prove mutually exclusive.[87] There a several goals, but only one instrument is available.[88] Whether regulation is seen as fair depends on the institutions that execute and control regulation.[89] The implementation of regulation itself hence causes failures and disturbs not only the market being regulated but others as well. Additionally regulation per-

---

[82] See North, 1992, p. 19.
[83] See Voigt, 2002, p. 17.
[84] See Vogelsang/ Mitchell, 1997, p. 53.
[85] See Heng, 2003.
[86] See Soltwedel et al., 1986, pp. 2 ff.
[87] See Eberlein/ Grande, p. 43.
[88] See Newbery, 2000, p. 2.
[89] See Rawls, 1971, pp. 3 ff.; Vogelsang, 2002. p. 5.

sists and once it is there is hardly ever completely abolished. That is why regulation is often called a second best solution.[90]

To find the right regulation for the particular market situation is a process of taking gradual steps so as to increase or decrease the amount of regulation.[91] As there are many different market constellations, there are also many different ways of regulating. In practice this causes a constant shift between regulation and deregulation.[92] Altogether regulation poses one of the central questions of industrial policy[93] since it is questionable whether efficiency goals, for example efficient negotiations, can be met through regulation at all.[94]

Résumé: Perceived power in the framework dimension. The rules of the game, i.e. property rights based on values, deliver security to those at the negotiating table:[95] actors know what to expect when they follow the rules and also what happens if they do not follow the rules. This is the opportunity that comes with rules and institutions. Rules also work as a restriction, though. It matters who sets these rules and who has access to them. It could be that those who do not have as much power as others in a negotiation will not be as successful because of the rules. The way property rights are distributed and institutions are established can be seen as a form of discrimination against those who do not have the opportunity to set the rules. On the whole regulation is the 'mother' of asymmetrical distribution of power based on the framework. This also demonstrates that regulation is a prominent example of the need to negotiate on different levels. The existence of power based on the framework has been shown. What the exact effect of this power base on negotiations is, remains to be discussed.

This leads to **Hypothesis 1**: *Property rights are distributed asymmetrically between the actors and as such are a source of perceived power in negotiations.*

Trying to bridge uncertainty causes dynamics of power

Actors might want to react to the framework that surrounds them. Some of this kind of action is limited by the framework and some of it is enhanced.

---

[90] See Kahn, 1988.
[91] See Engel, 2002b.
[92] See Bundesregierung, 2002c.
[93] See Armstrong/ Cowan/ Vickers, 1994.
[94] See Soltwedel et al., 1986, p. 16.
[95] This paragraph builds on North, 1992, pp. 58 ff.

The following section will analyse who the actors of a negotiation are and what it means if they act or fail to act rationally. It is argued that the theory of public choice can work as an indicator in measuring the dynamics of power in negotiations. The dynamics of power are set in motion on different levels, and the actors try to limit but at the same time cause them.

Who are the actors and on which level are they active? It is obvious who the actors at the negotiating table are. However it is not only those at the table who are part of the negotiation process. Even though theory often looks only at those actors who are directly involved, very often there are more 'things happening' behind the scenes. This is the case, for example, when a politician issues a statement regarding a particular negotiation. But not all such interventions take place openly. Lobbying and influencing might happen 'under cover'. The actors might, in addition, have formal or informal connections to each other or to other actors in the background. It can be concluded that the setting within which the negotiations take place includes more than those at the table.

The composition and the size of the group is, then, obviously one important parameter in the analysis of negotiations and of power. One could say that negotiations take place on different levels: the endogenous level allows us to look only at the actors at the table of a negotiation, whereas the exogenous level also includes those that do not directly take part in a negotiation, but still influence the process and possibly the result. It is not clear if a negotiation takes place only on one level or on several at the same time. It is thus not always clear who is actually involved in any given negotiation. One way or another the actors will explore their possibilities within the framework, but also outside the framework in order to reach the best result in their individual negotiation situation. They will try to discover in doing so the boundaries of the negotiation. Once an actor has explored his possibilities within the network he can make a decision between doing nothing to change the framework and doing something to change the framework in order to influence the negotiation. If the actor decides to 'do something' he still has the choice of different levels and intensities of activity. The cost of changing rules is different (asymmetrical) for different groups of actors.[96] Obviously, different actors have different kinds of access to the various negotiation levels. Depending on the way the framework is set up, it might restrict the actor or it might offer him opportunities to use power by which to change the framework itself.

Actors have only a limited ability to process information. Economics assume the actor will behave in a rational way. If this were indeed the

---

[96] See Buchanan, 1962, p. 350.

case, then it should be easy to predict the outcome of a negotiation and there would be no difference between perceived, resulting and real power. But in reality actors act 'irrationally', or so it seems: there is a 'maze' of different interpretations.[97]

Irrational behaviour might be caused by the limited ability of the actors to process information.[98] Such a limitation is called bounded rationality: even if actors were to have access to complete information they would not be able to make decisions that are based on every detail of it.[99] Considering that actors are not able to process all the information which is available to them and relevant for the decision-making process, every contract will in some respect be incomplete. This means that the actor also has incomplete information about his options. The actor is limited in his choice of actions as he cannot picture all the possibilities that are available to him. He is not able to process all the information which would enable him to decide what is important and what is not. Still, actors need and want to reach an agreement, which is the logical reason for their entering into negotiations in the first place. To cope with their limited processing ability actors develop, based on their own values, a 'normal' reaction in negotiation situations. 'Normal' can be translated as rational behaviour in economic theory.[100] The limits of the concept of rationality can easily be explored by 'reframing'. Reframing means asking a question with the same content in different ways. It can be shown that reframing brings different results and can thus not be ignored as a side effect which only affects a minority.[101] It is, rather, an effect which is fundamental to almost every negotiation. Actors make use of the references to hand and arrive at 'irrational' decisions because of the phenomenon of overconfidence, because of a certainty effect and because of loss aversion.[102]

Man is not fully rational,[103] but limited rationality in negotiations is what seems to be closer to reality.[104] Actors choose to look 'only' at those parts that are within their perspective. This is perhaps not enough, on the one hand, to achieve the best possible result, but it could be the most efficient way of handling negotiations: narrowing one's view in order to obviate the cost of obtaining an overview. An extreme step would be to focus on only

---

[97] See Simon, 1982, p. 417.
[98] See Simon, 1957, pp. 241 ff.
[99] See Simon, 1982, pp. 409 ff.
[100] See Pen, 1952, p. 29.
[101] See Tversky/ Kahneman, 1986, p. 722; Bazerman, 1985.
[102] See Kahneman/ Tversky, 1991, p. 4.
[103] See Coase, 1984, p. 281.
[104] See Voigt, 2002, p. 29.

one level, hoping that this will deliver the desired result. Simplifications are a normal stratagem if one is not able to handle complex situations.[105] Simplifying the decisions by 'shutting off' certain inputs or knowledge, enables an actor to focus on the 'important' decisions. Such a focus will be hard to create artificially, though.

Young considers it necessary to use philosophy as a basis for today's 'economic rationality', including the culture and the way people think and react.[106] This is the 'philosophical setting' or the framework within which the actors agitate. Rationality is '*Wohlbegründetheit*', which translates as "the capacity to give good reasons for beliefs and actions".[107] Good reason helps the actor to be rational, because the framework lets him react quickly, almost reflexively. The framework is included in such a decision. It is based on three philosophical ideas about human behaviour, which have been translated into economic theory in the past: egoism, practical reason, and a belief that the end justifies the means. In this respect the actor becomes rational only with experience, having learned from the environment.[108] This would also be to recognise that 'learning by doing' is a necessary step in building a relationship in negotiations. Learning implies a time lag, though, since learning requires time for reflection.

Public choice as an indicator of the dynamics in negotiations. The actors at the negotiating table have been extensively described. More stakeholders have to be included explicitly in this picture.[109] Public choice could be appropriate for this task, as it analyses the integration of (political) man in the economic market place. "Public choice can be defined as the economic study of non-market decision-making, or simply the application of economics to political science."[110] The same principles that are valid for the market place are applied to actors in the public arena and the questions being asked in public choice are the same as in economic theory: which preferences do the actors have, can an equilibrium be found, and what is the Pareto efficient solution? The theory of public choice thus expands the economic standards and methods that are used to describe business processes and to explain (human) interrelationships. Public choice assumes that actors behave rationally and that they try to maximize their utility, just as economic man does. The utility function can include, depending on the actor, not only maximum profit, but other 'soft' parameters as well. Eco-

---

[105] See Simon, 1957, p. 242.
[106] This paragraph builds on Young, 2001, pp. 3 ff.
[107] Young, 2001, p. 11.
[108] See Conrath, 1970, pp. 195 ff.
[109] See North, 1992, p. 81.
[110] This paragraph builds on Mueller, 2003, pp. 1 ff.; Mc Nutt, 1996.

nomic theory is hence no longer limited to picturing the pure market-economic players, but also includes those that define the framework and the institutional setting, i.e. all the participants. Values are expressed in terms of utilities and rational economic behaviour.

A dynamic analysis is needed to help describe interaction between the different actors (on the different levels) and thus delivers a more realistic picture of a negotiation.[111] Public choice offers the theoretical framework[112] for doing so by enlarging the 'narrow view' of negotiations. From the outside the perspective is enlarged so that not only the actors of the negotiation but also those surrounding the negotiation become part of the picture. Public choice allows the inclusion of all the actors and backgrounds in order to see who the participants are and on which levels they manipulate the negotiation to reach their goals. Additional stakeholders could, for example, be public officials issuing ordinances. The actors could try to influence these institutions.[113] Public choice shows that it is rational for an actor to focus not only on the negotiation itself, but also on the framework that influences the institutional setting. In that way public choice is an instrument revealing the connections between the visible and invisible actors.

Decision-making is at the core of the political process.[114] The political process is viewed as a means of deciding what is right and wrong.[115] Policy is set by a fairly small group, if not an elite, of scientists, economists and civil servants in ministries. They do not stand for a single idea, but contradict each other heavily in their opinions. The parliament has hardly any direct influence over them, because many members of parliament do not have the necessary background to discuss the wide variety of issues that they are confronted with on a daily basis. Many of the public officers try to implement the measures they consider to be right, but there is hardly any coordination between them. On top of this, public choice indicates that the participants are unlikely to show transparently what their interests are.[116] Even if they do announce their goals and interests there is no way of proving that what they announce is true. Trying to influence the framework is then primarily political work. Politics make the job of influencing costly, though. There is a great danger that the framework, when it becomes a 'political football', causes costs in the form of political bewilderment, which can only with great effort be expressed as a cost function.

---

[111] See Simon, 1966, p. 17.
[112] See Vaubel/ Willet, 1991, p. 93.
[113] See Stigler, 1971; Armstrong/ Cowan/ Vickers, 1994.
[114] See Simon, 1966, p. 15.
[115] See Buchanan, 1966, p. 26.
[116] See Mueller, 2003, p. 160.

Public choice expands the perspective on negotiations in two ways and at the same time functions as the glue between the different parts of the theory. Firstly, public choice 'makes room' for those actors who do not actively take part in the negotiation. Institutions are also allowed onto the playing field by public choice, even though "the interrelationship between the various elements is often subtle and intricate."[117] Secondly, public choice can show how different levels of participation in a negotiation can be fine tuned so as to work as an instrument. Public choice provides the opportunity of understanding, or at least rating, the behaviour of the actors in negotiations by using the framework as a scale. The framework and the theory of public choice are of help as a point of reference, securing the actors' own objective standard,[118] but at the same time helping to support rational decisions. Public choice might help to understand better how rationality in negotiation situations works; it could help to set up an indicator for dynamics of power in negotiations. The efficiency measure consists of discovering whether or not changes are possible and if so at what cost. Also, the soft factors, such as values and emotions, can be expressed in economic terms so that an economic view of the negotiation becomes possible.

Capture demonstrates that regulation can become an end in itself. As public choice allows one to include soft factors in the analysis, it sets up a bridge to the theory of public goods and regulation. Markets fail to provide certain goods. This sometimes makes it necessary to provide them as public goods, because the individual does not have the appropriate incentive to provide them on a private basis. To find the right balance between the provision of private and public goods is not easy. Contingencies in contracts make a framework necessary for negotiations and government intervention, i.e. regulation, but even such a framework is incomplete and does not come without cost.[119] For example frameworks create regulation which in turn creates institutions that develop a life of their own, which is costly. Even worse, once a market is regulated it is not easy to de-regulate it again; instead regulation becomes an end in itself:[120] this is known as capture. Regulation can thus grow to become the only reason for stakeholders to exist, as regulators are interested in securing their own existence over the long-term.[121] The result might be that they are not necessarily keen on helping in the actual negotiation, but rather on making sure that their own

---

[117] Mueller, 1989, p. 6.
[118] See Baumol, 1986, p. 9.
[119] See Furubotn/ Richter, 1997, p. 21.
[120] See Stigler, 1971.
[121] See Mueller, 2003, p. 359.

power and jobs remain secure.[122] Conflicting interests are the basis of this kind of relationship. This points to a close connection between the theories of public goods, regulation and public choice, as public choice connects the bureaucratic aspects of the framework with the economic behaviour of the actors. The existence of regulation can also be explained by the work of interest groups.[123]

The dynamics of power are set in motion on an institutional and on a 'personal' level. In the context of this study the question is, when, where and how will an actor try to change the framework? Where is the source of power 'attacked' and the dynamic of power applied? Actors do so on the institutional and on the personal level.

Actors try to apply dynamics to the **institutional level**, to regulation as such, in order to test its limits. The framework within which the actors operate in one way restricts but in another way offers possibilities for the scope of negotiations. As there is a co-existence of formal and informal rules, both incomplete to some extent, the search for a more complete institutional arrangement is the cause for dynamics and change. As laws do not cover every aspect and detail of the situations that can arise, they act like a guide. This way there is always room to bend laws, to uncover situations or constellations that are not fully covered by law. The concepts of limited and bound rationality make it obvious that information is invariably missing and disagreement between actors is to be expected.[124] The key terms regulation, liberalisation[125], de-regulation and re-regulation[126] describe a dynamic change that takes place over time – where, for example, with the help of the framework a sector is being developed from a monopoly to a competitive market.[127] But such an environment of change often leads to conflicting goals between the different levels of the framework.[128] In this sense regulation is a dynamic element in negotiations which surrogates the pressure that otherwise competitive forces would exert in order to deliver economic efficiency.[129] Regimes which cause asymmetrical negotiation situations for the actors will probably only work for a limited time, though.[130] In this sense regulation can be a reason for negotiations – trying

---

[122] See Pesch, 2003.
[123] See Soltwedel et al., 1986, p. 4.
[124] See Furubotn/ Richter, 1997, pp. 19 ff.
[125] See Pelkmans, 2001, p. 439.
[126] See Wegmann, 2001, pp. 50 ff.
[127] See Newbery, 2000, p. 399.
[128] See Héritier, 2001a, p. 2.
[129] See Kiessling/ Blondeel, 1998, p. 571.
[130] See Lyon/ Huang, 2002.

to ease the burden of regulation by defining one's own rules. This signifies the acceptance of the fact that the regulator works as a filter and it is not obvious for the actors what exactly passes through it. In the absence of a functioning market system the goal is to imitate the market mechanisms, but still any intervention can only be regarded as a second best approach.[131] There are, obviously, many different degrees of regulation which provide the actors with more or less incentive to change their situation, be it via negotiations or other techniques. Finding the 'right' mechanism and the 'right' time to use, to implement and to withdraw regulation is the challenge which the actors face. A built-in mechanism which depicts the development of regulations and, for example, lets regulation run out after a certain time or when certain criteria are met, would help to make regulation and its effects more visible for the actors.[132]

Actors also try to apply dynamics with reference to a more **'personal'** interaction level. It is evident that the interaction between the regulator and the regulatee is important in the context of negotiations.[133] Both have to rely on each other, as the firm offers expertise and the regulator grants decisions. Depending on how regulators on different levels and the firms are organised, this exchange decides how much power each actor has to hand. The main question is on which level the actors best work together. Those that offer more, get more. In the context of network markets this is translated into the fact that those firms which have more expertise and information can profit more from regulation; incumbents have more expertise and can hence profit more from such an exchange than newcomers. On the other hand if several regulators exist to regulate only one market, the regulatees have more strategic levers to negotiate regulatory action down to a minimum. A situation where regulation is not the same for all operators but differentiates between the actors could be the original reason for asymmetrical power situations. Such a minority protection is set up to protect (just) one company or a special market-segment. The actor that is equipped with less property rights could aim to get such a 'minority' protection, i.e. an asymmetrical regulation. As the framework protects one market or one single firm it might endanger others. When entering a negotiation the actors have to decide whether they will accept the given property rights distribution and if they are to stay on this one level. It could be worthwhile trying to change the result by exercising influence at a different level. Establishing more regulation in this respect can be the cause of more uncertainty, as even more 'grey' areas are established. Some actors

---

[131] See Better Regulation Task Force, 2004, pp. 1-2.
[132] See Freytag/ Winkler, 2004.
[133] This paragraph builds on Héritier, 2001b.

may even use legal uncertainty as an instrument, in order to provoke action by the other actors: regulation offers security for all involved, but at the same time offers opportunities of breach of contract.[134] In these cases the actor has to make a decision as to whether to accept or to try to change his situation. Actors could try to evade the pure regulation approach by agreeing to work on the basis of 'self-obligations',[135] thus minimising the influence of the regulator and intensifying their own input.

Actors try to limit, but also cause dynamics in negotiations. Actors are aware of the dynamics in negotiation situations and they try to set up a 'guideline' to handle and possibly limit these dynamics. Depending on the individual situation they are in actors search for an optimal trade-off between regulation and negotiated agreements.

There is a willingness on the part of the actors to make use of bureaucracy and a framework, if only as a security-net in case something goes wrong. The actors strive for procedures that give them some kind of planning security. Within a negotiation situation the framework for the negotiations is mostly established in the beginning. Intentionally and non-intentionally a climate for the whole setting will be generated. Actors seem to expect asymmetrical situations in their negotiation situation, but without knowing who will be better off: insecurity is part of the system of a regulatory regime and also for negotiation situations. A framework, however, will never be able to fill in all the missing links and gaps, the contingencies of contracts. Not even regulators have the knowledge necessary to be able to rule on every single detail and therefore actors settle for contingent contracts.[136] Such contracts are one reaction to the need to cope with bounded rationality. In a certain sense they leave out issues intentionally. Contingent contracts respect reality insofar as they accept the missing information and the resulting limited rationality of the actors as a fact, since contracts can never be complete and no one combination of contracts can fill all the gaps. There will always be grey areas that are filled by the actions of the different actors. But actors search for an indicator and are even willing to invest in order to understand, to test, bend or even extend the limits of the framework. All of this is to enable them to evaluate the power situation in their negotiation, to create their own rationality. Rationality together with the concept of *Wohlbegründetheit* in this sense is an instrument by which to create an indicator – 'practical rationality' – which makes it

---

[134] See Lyon/ Huang, 2002.
[135] See Voigt, 2002, p. 274.
[136] See Furubotn/ Richter, 1997, pp. 17 ff.

easier, possibly cheaper, and hence more efficient, for actors to perform in negotiations.

Actors have their own ways of handling complex situations. Whereas *Wohlbegründetheit* brings (even) more variables into the decision-making process and adds flesh and blood to the bones of rationality, rationality tries to limit the information to a minimum. The concepts are not necessarily easier to handle in theory, but they help the actors to cope with the complexity of reality. Actors make use of such schemes and translate them into actions in negotiations. It is most likely that actors act rationally in general, but there are particular situations where actors 'overreact' and/or act irrationally.[137] The 'weak' factors of a negotiation thus have a greater impact on the negotiation than one would have first expected: they act as an indicator that is able to identify (relatively) powerful elements within a negotiation showing if there is a lever that can change that situation.

Résumé: Resulting power in the framework dimension. The framework can assist the actor in deciding what he can do in respect to the use of power in negotiations and how it might affect the efficiency of his negotiation. Power plays an important role when a market is regulated. Regulation causes asymmetries.[138] Influencing the framework is political work in the first place: political influence on the framework is possible as there is no framework that is completely free from politics.[139] The framework is a rather long-term structure and results do not come about quickly. A benchmark which allows the actor to decide which approach to choose in a certain situation is only available ex post:[140] this is **resulting** power.

This leads to **Hypothesis 2**: *(Re-)actions of the actors lead to power-dynamics.*

Actors have access to different levels of a negotiation

The framework is both a source of power and a cause for dynamics in negotiations. By identifying the sources of **perceived** power within a framework, for example property rights that are distributed asymmetrically, it is theoretically possible for a negotiator to determine the outcome of a negotiation. This demonstrates that the framework is a source of power which the actors appreciate as a guideline for their negotiations.

---

[137] See Hauser, 2002, pp. 104 ff.
[138] See Perrucci/ Cimatoribus, 1997.
[139] See North, 1992, p. 81.
[140] See Kohlstedt, 2003b.

The common rules and settings, i.e. the framework, can be influenced even by the individual actor. Actors try to change what they sense as asymmetries. Even though actors have only a limited ability to process information and since there is no complete set of information for both, actors have ways of reacting within complex negotiation situations; but this will cause reactions by the other actors. Actors have an interest in reaching agreement but these interests and the actions directed to achieving the agreement are not necessarily identical.[141] These dynamic reactions lead to **resulting** power.

The framework turns out to be a restriction for a negotiation when the source and the dynamics of power are not harmonised. Actors search for an anchor within their accustomed environment using values as an index of power.[142] At the same time the actors might be using their knowledge about the habits and behaviour of others as an instrument, which in turn might lead to dynamics in the negotiation. Actions lead to even more actions and therefore risk the stability of the framework – i.e. what the framework was originally meant for. The dynamics of power are to be seen as a reaction to the sources of power, but even more to the actions of the actors. Perceived power is based on property rights and resulting power is caused by the dynamics of reactions. In one sense regulation itself is a reaction to the source of power, but at the same time it is the source of power for the negotiations which follow. As a dynamic reaction actors try to profit from the different levels of a negotiation. The connection between the behaviour that one is used to and how one reacts provides a link between the past and the future performance.[143] **Real** power, then, would involve making use of an indicator to enable the actors together to find the right time and access point or level at which to start a negotiation.

This leads to **Hypothesis 3**: *A problem cannot be solved on the level at which it originated.*[144]

The slow moving base of power, the framework and the reactions based on the limited processing abilities of the actors cause dynamics to develop when looking at and figuring out the grey areas, but can also create the opportunity to unchain innovation. In the case analysis it will be necessary on the framework dimension to identify both the sources of power – the key characteristic is the property rights distribution – and the causes of the dy-

---

[141] See Rubinstein, 1982, p. 97.
[142] See Simon, 1957, p. 64.
[143] See O'Connor/ Arnold, 2002, p. 8.
[144] Einstein, cited according to Schweizer, 2002, p. 221. Own translation from German to English.

namics which allow actors to access the different levels of a negotiation – the key characteristics are bounded rationality and incomplete information.

2.2.2 Cost forces actors to create options in negotiations

Attempting to lower the transaction costs is a common reason for negotiations. No individual can produce every product or service that is required to meet his needs. There is a good reason for this: to acquire the specific know-how necessary to produce everything for oneself is more expensive than trading the goods that one needs with those individuals who produce what they are best at producing. That is what the market mechanism is for: allocating scarce resources, so that best use is made of them for the good of the whole economy.

The specific nature of network markets could be the reason for certain extra **costs** in negotiations. For this reason firstly this section will focus on the resources that are employed in a negotiation and how they influence the results. How can an actor get the best from his individual situation in a negotiation? The principal agent theory and essential facilities will be discussed as a possible source of power. Secondly, cost is identified as a major driving force and as such a cause for dynamics in negotiations. The illusion of control and a possible indicator to determine one's own position in the negotiation are discovered. Finally, and as an initial résumé, the question as to whether actors are willing to invest to minimise uncertainty will be discussed.

Transaction cost and the access to essential facilities are a source of power

In this section it will be argued that negotiations are a part of transaction costs. It will be shown that different actors have different cost structures and different degrees of access to information. This kind of asymmetry creates 'big' and 'small' actors but also uncertainty.

Transaction costs cause uncertainty. Many economic models assume that the price mechanism, which helps the individual to make a decision as to where, what and which quantity to buy, does not cause cost. Coase was the first to point out that using the price mechanism entails a cost, even though many economic models take it for granted that such use is for free.[145] These costs are called transaction costs and are defined as the costs that occur when trading. But what exactly are the components of transac-

---

[145] See Coase, 1937.

tion costs? Coase names the identification of the actors, the formulation of one's wishes, the actual negotiation, the drawing up of a contract and the controlling of the contract.[146] Such an enumeration of the elements of the transaction costs does not solve the problem of quantifying the transaction cost, but it does help to identify the cost drivers. Transaction costs are caused by the actors trying to limit the uncertainty about a (possible) trade. Communication processes (i.e. negotiations) are necessary to overcome the lack of information and the asymmetrical distribution of information and rights between the actors.[147] As each negotiation situation is different, the costs of negotiations vary with the specific situation and with different contractual partners. For example, actors do not only invest in order to limit uncertainty, but also in the 'production' of power in negotiations. The theory of transaction costs hence provides a link between power and transaction costs.

There is not one set price which can be read off the tag,[148] and which would give the actor an idea of how much the price mechanism is costing him, though partly this is the case since trade also saves costs in many cases, compared to self-production. Thus there is no homogenous act of using the price mechanism.[149] The individual has to weigh possible costs against the gains that the trade may produce. In economic terms the individual weighs the cost of the actual goods or services traded plus the transaction costs versus the cost of not trading. If the cost of a certain trade is too high, the individual will respond by not trading and will search for an alternative. If the cost of the trade is low, then the trade is likely to take place. The seller will only sell if he will be better off than he would be if he did not sell. Since all of the actors involved in the trade calculate this way, all actors should be better off after the trade than before – otherwise they would not execute such a deal.[150] Together with the price mechanism, trade becomes the instrument to find the balance between supply and demand.[151] Even in situations where 'small' actors are confronted with having to negotiate with 'big' actors the goal is still to lower the transaction cost. It might be necessary to be of a minimum 'size' or to have a certain amount of power in order to be able to enter into negotiations at all. If both

---

[146] See Coase, 1960, p. 114.
[147] See Mandewirth, 1997, pp. 36 ff.
[148] See Pen, 1952, p. 24.
[149] See Coase, 1988, p. 45.
[150] See Donges/ Freytag, 2004, p. 182.
[151] See Coase, 1988, p. 4.

actors are better off, they have reached a win-win situation[152] and have moved from a non-Pareto state towards a Pareto optimum. It can hence be concluded that transaction costs influence the decision-making process of the individuals regarding trade.[153]

The basis for every negotiation lies in the theory of transaction costs as discussed by Coase.[154] An actor, trying to minimise the cost of the trade, has an interest in minimising the costs of each part of his transaction.[155] Negotiations are one essential part of the transaction costs and conversely the existence of transaction costs is the reason for negotiations. If an actor can envision that by entering into negotiations the cost of the trade as a whole can be reduced, he will do so. One way to avoid or at least to lower transaction costs is to internalize them. Coase argues that instead of trading with external partners, one can set up a firm and get better results, because transaction costs for working with others would be saved.[156] Instead of trading, one would then set up a bundle of administrative procedures (which is basically what a firm does) for the internal production which then became necessary, which would not only include the communication between all processing levels but also that between production and sales. This set of administrative procedures is the basis of a firm in the world of Coase. Firms exist to lower transaction costs. But, obviously, the danger exists that organisational costs may increase (most likely disproportionately) with increasing company size. Up to a certain point, internalizing transaction costs is one way of lowering these costs. The existence of transactions costs thus causes some degree of uncertainty about how to best position oneself in a negotiation.

Principal agent relationships diffuse the power flow and create uncertainty. On the one hand actors 'become' individuals, distinct from the firm, with their own values during negotiations.[157] This explains why results not only depend on the contents but also on the 'bargaining ability' of the actors and on their personal reactions in certain situations. On the other hand an actor is not necessarily part of the negotiation as an individual, but possibly as an agent for his company or an institution. Principal agent the-

---

[152] For an extensive description of what is meant with a win-win situation see Hauser, 2002, pp. 33 ff.
[153] See Aufderheide, 2000, p. 154.
[154] See Coase, 1937.
[155] See Coase, 1988, p. 39.
[156] See Coase, 1988, p. 7.
[157] See Hauser, 2002, p. 50.

ory describes this relationship between the agent and the principal, which leads to two main problems that are a common part of any negotiation:[158]

Firstly, the agent may not stick to the course assigned by the principal. There could be different reasons for this. It could be intentional, as the agent might have or might develop goals of his own, for example because the individual does not share the same interests as his firm does (so-called hidden action). It could also be a misunderstanding insofar as the agent did not fully understand what the principal asked him to do; he might not know, for example, what the reason for a certain line of argumentation is. The behaviour of the agent causes the need for the principal to 'protect' himself against such a course of action and hence incurs a cost. Secondly, more actors make negotiations more complex as the demand for harmonisation between the different layers (or hierarchies) of the firm becomes necessary. This need for harmonisation might make it necessary to implement certain modes of control between the principal and the agent. Much of the power then goes into the internal process of coordination (so-called hidden information) instead of the original cause for the negotiation. In that way the investment of the actors in power makes negotiations more complex and thus more costly.

Principal agent theory connects the aspects of access to information and the ability to process this information.[159] This may be expressed in the following economic terms: getting and handling information causes costs, but it is absolutely necessary for negotiations. In relation to power and negotiations it becomes obvious that information and the handling of information is power and will be used by the actors. Even the regulator-regulatee interaction mirrors this: the regulator is uncertain about the actions of the firm as the firm tries to make use of the grey areas that exist in the framework.[160] The firm tries to lower its transaction cost by making use of the informational asymmetry and the regulator by using his hierarchical control. On the cost level information is therefore a bottleneck which poses an economic problem for the actors: how to get it with the least cost whilst running negotiations efficiently. Both the principal and the agent accept that the relationship causes cost, but they are ready to incur these extra cost anyway:[161] they expect a positive cost-benefit relation out of the total result of the relationship. Thus transaction costs vary in different negotiation situations as the principal agent relationships differ as well. The diffe-

---

[158] This section builds mainly on Erlei/ Leschke/ Sauerland, 1999, pp. 74 ff. and Furubotn/ Richter, 1997, pp. 148 ff.
[159] See Erlei/ Leschke/ Sauerland, 1999, p. 74.
[160] See Héritier, 2001b, pp. 3 ff.
[161] See Furubotn/ Richter, 1997, p. 151.

rences in the cost situation are caused by the different levels of access to information. Not knowing the cost situation in a negotiation is a source for uncertainty.

Cost of and access to information as an essential facility. One of the reasons why negotiations are so complex is the amount of information that needs to be handled. It is not costless to make the information transparent and accessible, i.e. exploitable for the negotiation situation. The actors even depend on each other to utilize the information, but different actors have different degrees of access to information; the accessibility of information might even depend on the size of the actor. In this way information might be the cause of asymmetries and hence a source of power.[162] Cost of information is the reason that principal agent situations in negotiations are part of everyday life and possibly one of the major reasons why negotiations take place at all. The advantage that an actor derives from being 'big' could be that he has something – for example information – that the other depends upon. Actors have different levels of access to certain resources which are deemed scarce: because of this there is a bottleneck. If there is no alternative to the bottleneck and if at the same time there is no reasonable way of duplicating the bottleneck then the conditions for an 'essential facility' are met.[163] The so-called essential facilities doctrine is about making sure that the resources can be used in an efficient way by the whole society.[164] This is fairly complex since that which is an essential facility today, is not necessarily one tomorrow. Depending on what is essential and what access is available to the actors, a need for negotiations emerges so that some or all of the actors may get or defend access to these scarce facilities. In the economic literature the characteristics of the essential facilities, i.e. the high amount of fixed costs, the economies of scale and the danger of ruinous competition are the basis for the so-called market failure doctrine.[165] Since actors are different, have different degrees of access to different resources and behave differently, the sources of power are perceived to be distributed asymmetrically between the actors.

The difference between 'big' and 'small' actors. There is a difference between the know-how possessed by an individual and the know-how that exists within a company.[166] This difference can be expressed in terms of cost, based on the possibility of access to the information available to an

---

[162] See Donges/ Freytag, 2004, pp. 186 ff.
[163] See Knieps, 2004, p. 20.
[164] See Engel, 2002b; Kirchner, 2003, p. 7.
[165] See Deregulierungskommission, 1991; Klatt, 1994.
[166] See Erlei, 1998, p. 173.

actor. For an individual (normally a 'small' actor) it might be easier to run through a quick decision-making process as there is less need to harmonise different internal opinions. The individual negotiator organises himself. The theory of transaction cost sees the behaviour of the individual as an important part of the bargain.[167] The bigger the actor, the more difficult it is for him to maintain power in all situations, but at the same time he has more possibilities to do so.[168] Smaller actors profit from being more flexible. Negotiations seem to become not only more complicated with the use of power, but also more unstable. Unstable situations might cause uncertainty and inefficiencies in the negotiation. This points to a dynamic circle of actions and reactions, which prevents the actor from clearly positioning himself in a negotiation. The individual has a better opportunity than a firm does of delivering input for the creative process of finding options, as he does not need to go through a hierarchy getting permission to make decisions. The difference between 'big' and 'small' thus seems to depend mainly on the number of hierarchies.[169] The more hierarchies there are, the more complex and more costly it will be to get the information in the way one needs it.[170] This deserves comparison with the effects of bureaucracy vs. individuals described above. Drawing such a distinction between firms and individuals in negotiations is important in order to demonstrate the different effects that 'big' and 'small' actors have on power in a negotiation. One could summarise this effect by stating that the rational behaviour of individuals is different from that of companies.

Résumé: Perceived power in the cost dimension. In a world of 'big' and 'small' actors, principal agent behaviour and the essential facility information are a source of uncertainty and therefore **perceived** power in negotiations. The distribution of information, possibly based on the framework, constitutes the difference in power between the actors. The cost of power and the cost of negotiations in that sense are the cost of access to and of handling information.

This leads to **Hypothesis 4**: *Transaction costs are a source of uncertainty.*

---

[167] See Williamson, 1993, p. 5.
[168] See Héritier, 2001b, p. 20.
[169] See Donges/ Freytag, 2004, p. 249.
[170] See Erlei, 1998, p. 172.

Uncertainty creates dynamics of power and dynamics create uncertainty

Scarce resources make it necessary for actors to (re-)act. Negotiations, the heart of the trade, are one such way of (re-)acting in order to ensure the efficient allocation of resources. When handled correctly, all actors involved will be better off than they were before and will be certain about how the trade can best be accomplished. Trade and transaction costs are part of the market system, and are needed to find the balance between demand and supply. Negotiations are a process of dynamic moves on the part of the actors and these dynamics cause change.[171]

The question, then, is why actors invest in power and thus cause dynamics – and for what reason. This will be discussed in the following section under the heading "The illusion of control"[172], which means that actors are trying to control the uncontrollable by using power. The reason for such an investment is the uncertainty arising from information deficits. If there were some way of measuring the cost of power and its results, actors would be able to invest efficiently in negotiations. As such mechanisms do not really exist there must be a relative scale which allows one to compare alternatives.

'The illusion of control'. Actors try to exercise total control over the negotiation. It seems as if actors get satisfaction "from being able to control the seemingly uncontrollable."[173] The goal is to obtain certainty about the result of a negotiation as early as possible, preferably in advance of its commencement. But by doing so actors seem to only evaluate a given moment, without regard for their overall long term situation, which can violate logic theory and cause irrational results.[174] As actors focus on different time-horizons for their ongoing negotiation, the originally identical priority settings of both actors suddenly change to become non-identical points between the actors. As a result they might switch from the first of their original priorities to their second,[175] as negotiations do not only consist of one single issue. When trying to control a limited amount of resources,[176] cost reduction is handled by the alteration of issues.[177] Such a change of mind is not fully transparent to all the actors involved in a nego-

---

[171] See Furubotn/ Richter, 1997, p. 23.
[172] Langer, 1982.
[173] Langer, 1982, p. 238.
[174] See Kahneman, 2000, p. 707.
[175] See Hart, 1976, p. 300.
[176] See Zartman/ Rubin, 2000, p. 10.
[177] See Emerson, 1962, p. 34.

tiation situation. Non-transparency then creates a disturbance in the communication process, leading to uncertainty.

In constructed test scenarios it has been shown that actors behave differently, depending on what they have been told about the situation that they are in, even if it made no objective difference to the subject of the negotiation.[178] This shows that communication is essential in negotiations. Even with a starter package of property rights, which seems to favour them, actors do not experience more stability with respect to the distribution of power between the actors.[179] It seems to be difficult for actors to picture themselves in a larger context of a negotiation as predictions of the outcome of a negotiation based on a certain input of power are impossible. Even the goal of setting up a system of probabilities to picture the results is unattainable.[180] Since there is not only one optimal solution, but many to choose from, it is likely that the actors will only be able to state ex post which choice would have been best for them in that particular (one-off) situation.

Actors do not seem to be aware of the long-term, multilevel situation that they are part of. As a result their appreciation of power is only momentary. It has been shown that negotiations are better conceptualized as interrelated exchanges, rather than separable incidents.[181] A paradox seems to be the result: on the one hand the actors seem to notice only the power that they have in a particular situation (and which they might be tempted to use), but on the other hand they are aware that this one-off situation has an effect in the long-run on the overall situation between the actors, as results show that past negotiation performance influences the quality of subsequent outcomes. No stable concept of power exists which is valid for all situations. If an actor is powerful at one moment, he might not even be interested in being the more powerful in all the negotiation situations to come. There is a constant shift between different situations with different actors and different subjects which makes it impossible to predict when and whom has more power.[182] "Normally actors have only partial control over events which are consequential to them."[183] Power is then not an attribute of a person, but power exists for the moment or the situation.[184] This again highlights the importance of time and content even for the dynamics

---

[178] See Druckman, 1971, p. 529.
[179] See Lewicki et al., 1994, p. 294.
[180] See Zartman, 1974, pp. 387 ff.
[181] See O'Connor/ Arnold, 2002, p. 2
[182] See Simon, 1957, pp. 66 ff.
[183] Hart, 1976, p. 297.
[184] See Emerson, 1962, p. 31.

of power on the cost-level: because of their uncertainty about the times when events take place, the actors start a loop of actions and reactions.

Cost factor uncertainty. The power of the moment as described above 'survives' because of the uncertainty, which is nothing else than insufficient information about time and content. Not having enough information starts the dynamics of power: actors are trying to compensate for the missing information by the use of power: knowledge is power. Uncertainty is a starting point for many conflicts and thus a cost factor in negotiations. Such uncertainty can be caused by information asymmetries. Power relationships between actors are influenced by the distribution of information between them. Actors try to cause information asymmetries on purpose, but just as often the latter are rooted in deficits of communication. This confirms the need for good communication, which is an integral part of bargaining,[185] in order to eliminate the uncertainty. This need is particularly pressing, since conflicts and uncertainty leave the actors with lingering emotions and cognitions. The more personal an issue gets, the more often power and instability will take over in negotiations, since actors will do almost anything to achieve their goal, even having recourse to the use of threats.[186] It is in this sense that it would seem that in negotiations, where the actor represents himself (as opposed to representing a company), there is more asymmetry than in situations where the actor represents others.[187] By behaving this way, actors create uncertainty.

The psychological positioning of the actors may cause them to spend more time on taking in a position than on finding a solution.[188] They are likely to become even more committed to their positions, encouraging a zero-sum (conflict-)negotiation.[189] The more (time) an actor has invested in a certain position the costlier it will be to convince himself to give up parts of his position for the good of both actors; as time passes, actors slide 'deeper' into their positions.[190] Actors are even willing to go along with a bad result when trying to protect their own position, just to obtain the certainty they strive for as quickly as possible. Trying to save time is one of the major reasons why actors negotiate,[191] but building up power also takes time. The cost-benefit of their actions does not seem to matter in the short run any longer. One reason for this could be the absence of an obvious in-

---

[185] See Cross, 1969, p. 5.
[186] See Williamson, 1993, p. 6.
[187] See Druckman, 1971, p. 531.
[188] See Breidenbach, 1995, pp. 107 ff.
[189] See O'Connor/ Arnold, 2002, p. 5.
[190] See Deutsch/ Krauss, 1962, p. 74.
[191] See Cross, 1969, p. 12.

dicator. Thus, trying to measure this inability to grasp the result of the negotiation is an important part of the theory of transaction cost.[192]

Negotiations are one way of lowering the cost of uncertainty by closing information gaps via communication with the other actors. Studies have shown that close relationships lead to more integrative outcomes.[193] The benefits of a successful negotiation arise through reduced transaction costs for both actors. Reduced transaction costs for all actors raise the likelihood of a successful negotiation. It is therefore not necessarily just the use of power in negotiations that leads to positive results. On the one hand it could actually be that the investment in the production of power causes the opposite of the desired effect: the less power produced, the 'better' the negotiation result. It could hence be more efficient for the actors to invest in better communication in negotiations instead of trying to strive for a level playing field in terms of power distribution between the actors. On the other hand, the very existence of power might lead to good results as some people may work better under pressure. Power makes people negotiate and hence can add value to a particular outcome. Without the involvement of power, the outcome could be a less valuable result.[194] It is the uncertainty of how actors react to the use of power and what power does to a negotiation situation that causes costs for all actors but no measurable negotiation results.

Cost of power production. Uncertainty is not the only reason that power is costly. Using the instruments of power and establishing a power-position, keeping up with the role an actor plays, involves the use of resources. Thus power and cost are closely connected – but what is the cost of power?[195] The production of power needs to be looked at in terms of a cost benefit analysis: how much input of each factor is needed in order to obtain a certain result? Or, to put it another way, how does the investment needed to achieve a certain result relate to the investment in an alternative option? The key to using power in negotiations, then, seems to be knowing when to invest, how much to invest and in what kind of power. But there are hurdles in the way which prevent the actor from getting this certainty.

The theory of transaction costs shows that it is possible to lower them. This can, for example, be achieved by internalizing the transaction costs, as discussed above under the 'Coase Theorem', but also by investing in better mechanisms. Such a mechanism could be the implementation of power: as actors strive for more certainty they are willing to invest in

---

[192] See Williamson, 1993, p. 25.
[193] See Foster, 2002, pp. 5 ff.
[194] See Zartman/ Rubin, 2000, p. 11.
[195] See Zartman, 1974, p. 396.

power, thereby causing transaction costs. The cost of the 'wrong' use of power can make negotiations inefficient. But the restrictive nature of cost can be overturned when it is made to work as an instrument of power for those that implement it efficiently. The question seems not to be whether power is part of all stages of a negotiation.[196] The question should rather be: how can actors make use of their power and still go home with good results?[197] One possible answer could be that the actors have accomplished the setting up of a communication structure between them which allows them both to profit from their investments in power without having to cope with all the disadvantages of the use of power. Here the results again show a mixed picture when looking at firms and individuals, as negotiations seem to become only more complicated with the build up of structures within firms and the use of power. The cost of the negotiation process, or rather the cost of achieving the result of the negotiation, can in some cases be minimised by making use of power. Dynamics in power distribution are then caused during the search for better negotiation mechanisms by the actors. This also shows that dynamics are both the reason for and the cause of asymmetries: as one actor might have more resources and might even be willing to spend those on achieving a certain result in the negotiation, this could also endanger the whole process and thus also the certainty that is striven for. To obtain a clearer picture of the effects of power it is necessary to find some way of measuring the cost of power.

Measuring the investment in power. It might be necessary to analyse what an investment in power does to the negotiation in order to find out if the goal of (increased) certainty can be achieved more easily thereby, or to put it another way, to discover whether such an investment is efficient. Being more powerful could promise a better result in a negotiation situation. The more resources there are available to the actor the more he can invest in the 'production' of power. Power is a production function, where different amounts of resources are transformed into power.[198] 'Big' actors have more opportunities to exhibit and execute power, but if this power exercises undue influence the result of the negotiation will be uncertain, as discussed above. This leaves open the question as to how and into what kind of power the investment is transformed. If the amount of resources invested related exactly to the amount of power produced, € 100 would relate to 100 units of power and € 10 invested would relate to 10 units of power in the negotiation situation. The difference in the amount invested

---

[196] See Ury, 1992.
[197] See Zartman/Rubin, 2000, p. 3.
[198] See Harsanyi, 1962, p. 73.

by an actor in establishing and maintaining a position could lead to an asymmetrical distribution of power in the negotiation situation. But since 'small' actors have a different cost structure they might be able to produce the same amount of power with less resources – a process which might be more expensive, though.

This paradoxical uncertainty regarding the power production is one key to understanding the difference between perceived power and resulting power. The results of the power production cannot be measured (resulting power) as there are internal (for example principal agent effects) and external effects which are unpredictable (perceived power) and barely measurable. The principal sees only results, but does not know exactly how the agent achieved them.[199] It might be helpful to describe why it makes sense to produce cost in order to establish power in certain situations, and why it does not make sense to use power in other situations. The danger is that the investment creates and secures power only for a moment and possibly even for the wrong moment. This paradox seems to make it impossible to pick and measure the right investment in power. For this reason it seems to be necessary to find another kind of mechanism to compare different power investments in negotiations.

In every trade situation it is possible to reach not just a single solution, but several. It is most likely that the variety of solutions is the result of the negotiation-process of the actors involved in the trade.[200] "An agreement by its very existence is a good and has value in and of itself."[201] But there must be a way of expressing this value in more detail and deciding how the different alternatives can be rated. The question is, how can the results of a negotiation for the individual actor, but also for the group of actors, be measured? The two following concepts give an overview of the different possibilities when setting up scales: a cost benefit analysis and Pareto optimality as a relative concept, where the different options are weighed against each other.

A **cost benefit analysis** can help to compare the different power investments in negotiations: did the investment in power deliver the desired results and at what cost? Measuring power in a cost benefit analysis means taking into consideration the opportunity costs of actor A's attempting to influence actor B, and the opportunity costs to actor B of resisting actor A.[202] Step by step the possible advantages (for example, getting a better price) can be compared with the disadvantages (for example, causing a

---

[199] See Furubotn/ Richter, 1997, p. 22.
[200] See Coase, 1988, p. 105.
[201] Zartman, 1974, p. 387.
[202] See Harsanyi, 1962, p. 67.

conflict situation).[203] Any effect of power might be offset by the potentially (higher) costs of using power.[204] The use of resources in building up power versus the actual result is the measure of power in terms of efficiency. The difficulty again is that the real cost of a negotiation cannot be defined exactly, and thus the efficiency of it can only vaguely be calculated. Institutions, a framework and a starter package of property rights, can serve as an anchor in such situations. There is no complete institutional arrangement, though; this co-existence of formal and informal rules is a potential source of dynamics which creates uncertainty.[205]

In their search for certainty, actors do not necessarily know which rights they (or others) have, nor do they know if they have more power than someone else. Technical or individual efficiency means making the best use of scarce resources, i.e. avoiding their waste. Economic efficiency describes the situation where a given production is realised with minimal cost. Because quantifying the transaction costs of the power production is not easy, it is possible that a relative perspective of transaction costs will be helpful.[206] Such an approach must, however, allow for the dynamics of a negotiation. Welfare economics, whose perspective includes the whole economy, sees efficiency when the Pareto optimum has been realised. **Pareto optimality** is another measure of efficiency.[207] It is only when none of the actors involved can be made better off without making another actor worse off than before that a solution can be called optimal. Pareto optimality is a very strong criterion and can hardly be controlled by either of the actors, nor even by an external third actor. The Pareto criterion can (only) show whether or not one solution is an optimal solution, based on how the individual actor judges a certain solution. In a static model the Pareto criterion would be able to assign the best possible investment solution for all actors in a negotiation situation. Again the actor with more resources might find that even if he were to input an enormous amount of resources it would not help him achieve his desired result, as long as he failed to co-operate with the other actors. His powers would be diffused, even if he tried to aim for one certain result in the negotiation. Since Pareto optimality can only help in a static world, the dynamic concept of Pareto superiority could offer relief: Pareto superior results are better for all actors involved, compared to the alternatives that the actors have, but there is a

---

[203] See Pen, 1952, p. 32.
[204] See Coase, 1988, p. 24.
[205] See Furubotn/ Richter, 1997, pp. 25 ff.
[206] See Voigt, 2002, p. 88.
[207] See Buchanan, 1962.

chance that even better (theoretical) solutions exist.[208] If it is already difficult to measure the Pareto optimality, it will be even more complicated to calculate (or make a prediction as to) events so far unknown. This highlights again the paradox of power: ex ante knowledge is incapable of determining whether the perceived power is the same as the resulting power.

Actors depend on each other to create valuable options. Both of the concepts for measuring the efficiency of negotiations discussed hitherto establish benchmarks using different indicators. A cost benefit analysis is more limited, as the cost of both the transaction and the result of the negotiation must be known in detail in order to carry out the analysis. The concept of Pareto superiority offers the opportunity of comparing different options by way of indicators which are tailored to the individual negotiation situation. Such a relative concept also delivers a 'starting position' for a win-win approach on the part of the actors as it allows them to invent new options within the negotiation process, which could otherwise not be compared with the help of a cost benefit analysis. Both concepts show that the different actors can profit from each other with the right use of power when trying to achieve meaningful results in a negotiation. The bigger actor cannot turn 'his' result into an efficient solution if the smallest actor does not support this solution as well. Within 'big' companies, however, it might be more expensive to reach an acceptable position.[209]

It is the dynamic of power that causes cost in a complex world where actors are of different size and have access to different levels of information. Negotiations are one way of bridging these differences. To run them in an efficient way some kind of an indicator for measuring and determining the efficiency of the actions is necessary. A relative approach to measuring the degree of success promises the best results. Actors do not only compare the different settings within which they operate but they try to exert influence on them in order to get a better result. In so doing they compare different alternatives which they design in conjunction with the other actors.[210] This makes it obvious that no matter whether they are 'big' or 'small', the actors depend on each other to achieve a result in negotiations. The essential facility here is the limited nature of the information about the other actor and the consequences of the actions of both on the result.

On the one hand the actors might try to make use of the dependency which is created during a negotiation and contract 'making'. This effect is

---

[208] See Hauser, 2002, p. 21.
[209] See Buchanan, 1962.
[210] See Erlei/ Leschke/ Sauerland, 1999, pp. 273 ff.; Furubotn/ Richter, 1997, p. 32.

called 'hold up' and is caused by the bond that a contract sets up between the actors: as contracts are incomplete, actors have an incentive to make use of this dependency, from which the other actor cannot escape without cost.[211] If the cost of the 'hold up' is lower than the cost of escaping, the actor executing the 'hold up' uses power to his advantage and to the disadvantage of the other. This might have a negative effect, though, on the long term relationship. Again, as discussed in the framework section above, different settings might cause or inhibit such conduct. This pattern of thinking and acting in alternatives is the dynamic part of the principal agent theory. On the other hand the actors profit from differences in size. Actors depend on each other because resources are scarce and therefore they have a stronger urge to negotiate. In this sense negotiations cause dynamics which in turn generate innovative options of different possible deals whereby the resources might be shared.

Résumé: Resulting power in the cost dimension. Dynamics are one reason for the uncertainty which actors experience in negotiations and which are caused by the uncertain cost of power production and the non-transparent actions of companies and individuals. It is not necessarily advantageous to be 'big', as the complexity of an organisation increases with its size. The actors try to find an arrangement within which they can best handle the negotiation situation. As the relationship between the actors grows over time the complexity of the relationship might switch to a more routine behaviour, which needs less investment.[212] Thus it is necessary to look at the relationship of the actors on a time-line: there is an ex ante setting trying to get the best out of a situation and an ex post result, which is based on the setting but also caused by the dynamics of trying to change this setting when bridging the contingencies with action. This is what appears as an asymmetrical distribution of power between the actors in a negotiation situation.

This leads to **Hypothesis 5**: *Actors only have the illusion of control in negotiations as the dynamics create uncertainty.*

Non-transparency does not allow the translation of perceived power directly into resulting power. It would then be the dynamics which cause **resulting** power. The cost of the dynamics differs during the different phases of a negotiation, though.[213] To summarise this train of thought: dynamics are a mirror of the framework, the setting and the behaviour of the actors

---

[211] See Erlei/ Leschke/ Sauerland, 1999, pp. 182 ff.
[212] See Furubotn/ Richter, 1997, pp. 169 ff.
[213] See Erlei, 1998, pp. 153 ff.

in a negotiation and have an effect on the resulting cost. Negotiations cause results (resulting power) which can only be compared in relative terms. This leads to the need or the opportunity to develop new (innovative) options together.[214] Negotiations might offer a means of getting a grip on the power dynamics at the cost level, by trying to make the best of a situation for all involved, not just one actor.

Actors are ready to invest to create options in order to minimise uncertainty

Information is a scarce resource in negotiations. In some respects information is even an essential facility and a source of asymmetries. Principal agent theory and the advantages of long-term relationships hint at the advantages of self-regulation, i.e. agreements without the direct involvement of other institutions.[215] When the framework is secured, it could be the goal of the actors to find a solution without having to address third parties, as the best solutions can almost always be found between themselves.[216] Only if a market failure exists might institutional support, for example regulation, be necessary. Coase is sure that regulations "exist in order to reduce transaction costs".[217] Information is a source of **perceived** power but is at the same time a motivating factor which creates certain win-win situations in negotiations. One could say that negotiations are a way of handling the essential facility information efficiently.

Uncertainty is the main driving force with regard to cost aspects. Actors implement power so as to create certainty in their individual negotiation situation. In this way actors try to reach a state of symmetry by using power as an instrument. The concept of power for the moment demonstrates that the actors do not necessarily know where they are starting from and therefore might not be able to use their powers efficiently. It also shows that symmetrical and asymmetrical settings might be situated very close to each other: uncertainty does not allow for an exact differentiation of these two settings, though. The existence of transaction costs lead to negotiations, even in 'big vs. small' situations but at the same time seems to create even more dynamics and hence uncertainty. Cost is an instrument of power, but at the same time a possible indicator. Power is what the actors make of it and to what they compare the risk of losing. If they are risk averse they are likely to invest more in ensuring that they win. Thus the

---

[214] See Erlei/ Leschke/ Sauerland, 1999, p. 282.
[215] See Furubotn/ Richter, 1997.
[216] See Williamson, 1993, pp. 2 ff.
[217] Coase, 1988, p. 9.

individual attitude towards risk is both a cost factor and a scale.[218] Demonstrations of power are rather signs of perceived power than of real power. In the course of negotiations new (cost) aspects come up and need to be taken into consideration. By developing new options together, the Pareto frontier is moved and this gives all actors a new starting position. This move between positions is the shifting which the actors experience and which causes insecurities. The uncertainty is felt as uncertainty because the actors lose contact with their (original) point of reference, their starter package. This highlights the importance of the starter package, which affects the cost situation on the one-hand (based on perceived power) but also shows the diminishing effect of it in the course of the negotiation which leads to **resulting** power.

One of the paradoxes of power is that the making use of alternatives allows both the 'big' and the 'small' actors to find their specific advantages in negotiations. These advantages could be described as the **real power** in negotiations; but there is no real power of that kind which can help to predict results of negotiations. Actors, nevertheless, are ready to invest in instruments of power together to create more certainty by establishing options.

This leads to **Hypothesis 6**: *Negotiations can lower transaction cost even in situations of uncertainty.*

The key characteristics of negotiations with an asymmetrical distribution of power between the actors on the cost dimension that need to be borne in mind for the case studies are transaction cost and essential facilities as sources of power and uncertainty as a reason for dynamics of power.

2.2.3 Strategy makes transparency a favourable concept

How can an actor make the best of his individual negotiation situation? Most would intuitively say that he should make use of his strategic abilities. Strategy was originally a military term, describing the act of preparation and planning for a future situation.[219] In and of itself strategy is neither good nor bad. The opportunities for the use of strategy by one actor are most likely constituted by the limits of the other actor. Applying strategy in a negotiation thus means trying to make the best of one's available resources and information in the given context, with many unknowns still

---

[218] See Breidenbach, 1995, pp. 106 ff.
[219] See Lewicki et al., 1994, pp. 109 ff.

remaining. Altogether, so it seems, the success of a negotiation depends to a large extent on how it is planned.

In this chapter the key term **strategy** is developed on two different levels: firstly there will be a discussion of the ways in which strategy can help to set up a structure for a negotiation. This structure alone is a source of power in negotiations. Secondly, the application of the strategies by the actors is analysed as a cause for dynamics which could lead to more uncertainty in the negotiation. Game theory will be discussed as a possible tool and indicator of strategy.

Information and time are a source of power

Certain factors could be 'responsible' for the way results in negotiations are achieved when implementing strategy. A classification of the different types of disputes could be of help for the negotiators in preparing the strategy for a negotiation by breaking down a complex situation into smaller decision packages adapted to the individual situation. Possible classifications are here discussed under the headlines 'content' (as in information, i.e. quality and issues to be negotiated) and 'timing' (as in planning). These two classifications were already an integral part of the Power-Matrix in the other two dimensions, a fact which demonstrates the close connection of the three dimensions. The content aspect looks at what the actors put or refrain from putting on the agenda whereas the timing aspect describes when, how often and under what kind of time-pressure negotiations between the actors take place.

Which issues are put on the table? The importance of information. It is important to look at what is being put forward for negotiation. Just as important is to take a look at those issues that are **not** being discussed in a negotiation, though. The limiting or deliberate exclusion of issues can make it impossible to negotiate or to find options which suit all. Picking just one issue out of a mixture of many might cause results that no longer have any bearing on the original and main issue of the negotiation. Actors tend to design 'packages' of issues that they deem necessary to discuss in combination,[220] even if others do not see any connection between those issues. Even if all actors agreed to discuss only a single problem it would still not be certain that the negotiations would only revolve around this one issue. It can be shown that a negotiation of a seemingly single 'factual' issue implicitly involves further 'weak' issues such as emotions, values etc. An additional difficulty could be that the typical win-win situation invaria-

---

[220] See Hauser, 2002, p. 32.

bly requires opportunities for 'expanding' the scope of possible options. Focusing on one single issue only in a negotiation makes this impossible. Theory still tends to look at only one issue at a time, though, neglecting the point that interwoven issues are more likely to represent reality.

The more issues there are at hand, the more complex the negotiation becomes. There is a narrow path between having excessive and insufficient information and departing from this path will incur resultant costs. Determining the cost of information vs. the value of information for a negotiation situation is then the major challenge.[221] It could be of advantage for one actor to be able to limit discussions to a certain issue and to exclude others which are connected to the main issue, so as to be able to focus his energy more efficiently on the important parts.[222] This is making use of strategic power. At the same time the decision as to which issues are to be negotiated by the actors might limit their strategic possibilities.

When are issues put on the table? The importance of timing. Timing is a basic component of strategy: planning ahead for the things to come. One could say that strategy is an ex ante discipline. Timing has different aspects in negotiations. The number of rounds of negotiations are an important measure as negotiations are a learning process, i.e. they develop over time as the actors get to know each other better, so the more they get to know each other, the more efficient the rounds are likely to be. What do actors do to (not) let issues or discussions take place in time? Do the actors profit from first mover or second mover advantages and disadvantages?[223] This is making use of strategic power.

Theory often only looks at one-off negotiations, i.e. the actors will only negotiate once and no subsequent negotiation shall take place. Reality shows that these situations are very rare in economic life. Usually there will at least be a second round of negotiations (as the saying 'You always meet twice' reflects), even if this cannot be predicted at the time of the first negotiation. Most actors will thus bear the potential next meeting in mind when negotiating. Humans seem inclined to prepare for a potential long term relationship. Theoretical models tend to simplify matters and only consider one-off situations, i.e. using a static *ceteris paribus* approach.

Résumé: Perceived power in the strategy dimension. As the pattern for the negotiation is implemented ex ante access to information and the timing might be a source of **perceived** power and a cause of asymmetries.

---

[221] See Shapiro/ Varian, 1999, p. 3.
[222] See Hauser, 2002, p. 25.
[223] See Rieck, 2005.

This leads to **Hypothesis 7**: *The possibilities of reaching a positive agreement in a negotiation depend heavily on the access to information and the time phase.*

How the actors stand towards each other in the negotiation is influenced by the way the actors have strategically aligned these powers. Depending on the strategic setup actors can better control which, and at what time, issues are handled in a negotiation. Not only is this posturing with power already making use of strategy but at the same time it reveals the positioning of the actors in a negotiation, indicating whether they see themselves as weaker or more powerful than the other. Such posturing displays the relevance of perceived power to the actors. Already it is becoming evident that the structural approach of a strategy is decisive for a negotiation.

The actor's attempts to plan everything ex ante could place a restriction on the implementation of strategy, though. Actors try to obtain total control over the results of a negotiation by aligning the various instruments of power, without knowing exactly what their influence will be or what will happen next in the actual negotiation. This striving for 'security', because it is something which seems to steer the negotiation in the direction of a desired result, is valued highly by most actors. Thus, if the actors see an advantage based on the structural alignment of the strategic but only perceived powers on their side, they might be tempted to display them more than is efficient.

The decision to cooperate or not to cooperate creates dynamics of power

The actors are mutually dependent, but at the same time compete against each other in complex negotiation situations. On the one hand the availability of strategy can be an obstacle in negotiations as the threat of strategic actions by one actor constricts the actions of the other actors. On the other hand strategy could be the element of power that makes it possible for the actor to achieve whichever individual result in the negotiation he wants, provided he uses the right strategic elements. This already points to a possible paradox: the action of a single actor influences all the (other) actors, but is only caused by one. Each discrete action attracts the attention of another and is often countered by a reaction. What exactly causes what remains unclear – at least ex ante. As negotiators are confronted with multilevel and rather intricate problems, they want to be able to manage and understand these complex issues. They strive to react adequately, i.e. to make the best possible decision based on their (strategic-) possibilities. The dynamic environment of negotiations does not make it easier for the actors. For that reason they need some kind of tool which can assist them in such a situation. For the analyst such a tool is important in order to be able to understand the strategic level of power. Game theory describes the strategic economic theory of negotiations.[224] The most simple negotiation-game is a two person zero sum game, but more complicated win-win situations can be modelled as well.[225] Actors are thus offered an instrument for negotiations which helps them decide which strategies to use in order to reach the best results.[226] There are two major 'tasks' of the tool game theory with respect to negotiations.

Firstly, game theory is the scientific approach to describing the strategic dimensions of the actors in a way that can be applied by them: it helps negotiators to make a decision as to whether to cooperate or not to cooperate.[227] Secondly, game theory works as an indicator. Defining and implementing strategy means trying to be able to forecast reactions and results correctly. As all actors entering a negotiation expect an agreement that will leave themselves in a better position, without necessarily knowing what strategy the other actors will implement, it would be helpful to know what effect the execution of the different and independently defined strategies will have.[228]

Interdependencies shape the relationship of the actors. As strategy is about trying to steer a negotiation in a certain direction most actors expect to reach a certain result by adjusting their strategy.[229] To do so actors play for time or for speed, to allow or to prevent discussions from taking place, to make things happen or to ensure that certain things do not occur. Speeding up (cooperative behaviour) or putting 'a foot on the brake' (non-cooperative behaviour) thus define the speed of the negotiation. In due course actors make an active decision to negotiate either in a non-cooperative or in a cooperative manner. Game theory differentiates between non-cooperative and cooperative games and hence delivers a tool which supports an actor in a negotiation.[230]

Non-cooperative behaviour in negotiations is executed by blocking and threatening:[231] "To discourage entry ... play tough."[232] Whereas blocking is the threat to do nothing, threatening is the promise to do something bad.

---

[224] See Young, 2001, p. 21 and for the history of game theory see pp. 89 ff.
[225] See Nash, 1950, pp. 155 ff.
[226] See Young, 2001, p. 32.
[227] See Gibbons, 1992, p. 1 ff.
[228] See Bartos, 1967, p. 481; Allas/ Georgiades, 2001.
[229] See Harsayni, 1956, p. 145.
[230] See Lewicki et al., 1994, p. 38.
[231] See Harsayni, 1956, p. 144.
[232] Shapiro/ Varian, 1999, p. 31.

Blocking might be an interesting instrument for an actor who is obliged to negotiate but who has no time pressure placed on him: time works only to his advantage, as it does not matter when the issue is resolved. Threats are used by actors to speed up the negotiation process and possibly to push the implementation of a particular solution.[233] To execute a threat two actors are needed: the one who makes the threat and the one who is threatened. The actor that threatens aims to achieve his goal by way of a certain reaction on the part of the other. But different options are open to the one being threatened: he can submit to doing what he has been asked to do, he can defy the threat, ignore it or he can even issue a counter threat.[234] If someone is threatened by the same actor more than once he is likely to invest in establishing a better defence system. One could try to set up a framework to abolish threats or certain forms of behaviour. It is more likely that he will play along with his tormentor at first and seek some form of revenge later. It is important to notice that some kind of reaction is always awaited in response to a threat.[235] Even inactivity is sometimes regarded as an active response in such a situation.

To create uncertainty by setting up a threat system, the actor first has to establish his own credibility, though.[236] Credibility usually has positive connotations, but in this context it is only an indication that the threat could be put into effect. To make his threat convincing the actor must have at his disposal the opportunity of making the threats come true.[237] He can do so by making a promise. In this context a promise is a bilateral commitment, given a *quid pro quo*.[238] Power is then the ability to fulfil a promise,[239] offering the prospect of damage or asking for some sort of concession. The delivery of a promise works out even better if one actor depends on the other actor and his resources. Monopolies, for example, create dependencies, as they are the sole suppliers of certain goods.[240] If there is a state of mutual dependency between the bargainers, where there is no real substitute available and the other actor has the possibility of withholding the product or service this could be not only the sign that negotiations are needed, but also an indication that power is in the game.[241]

---

[233] See Schelling, 1958, pp. 203 ff.
[234] See Boulding, 1963, p. 428.
[235] See Emerson, 1962, pp. 36 ff.
[236] See Schelling, 1956, pp. 282 ff.
[237] See Külp, 1965, p. 38.
[238] See Schelling, 1958, p. 229.
[239] See Schelling, 1956, p. 299.
[240] See Berg, 1999.
[241] See Pen, 1952, p. 30.

If the other actor picks up on the threats, i.e. does what the threatening actor wants him to do, he will succeed in speeding up the process. This might even be to the (long-term) advantage of all actors by directly pointing to the best alternative available to all and therefore saving time.[242] However, if the other actor reacts less positively – with, for example, a counter threat – the negotiation might be prolonged. Counter threats seem to be paradoxical in nature:[243] actors do not do what they regard as best for them, but what is felt to be worst for the other actor. Threats are hence not about the interests of the actors, but about defending positions. In circumstances where uncertainty and threats thrive, the stability of the negotiation situation as such is constantly in danger.[244] Stability deteriorates further as bluffing, i.e. the deliberate misrepresentation of one's expectations in order to back up one's threat, is being used.[245] In business this seems to be an accepted form of behaviour, as long as certain moral standards are not breached.[246] Fooling and bluffing are, then, the intentional making use of uncertainty – pretending to have something which either allows one to make a threat or to create a counter-threat.[247] In this respect even lying may sometimes be an acceptable strategy.[248]

On the one hand threats are counterproductive and make it more difficult to reach an agreement,[249] as the use of power causes resistance.[250] Threats thus risk the relationship as a whole ('I cannot trust him any longer') and therefore lead to instability in the long run.[251] But on the other hand threatening behaviour is often a powerful social organiser and in this respect there exists a certain legitimacy of threat. One way or another threats are probably the most visible and open sign of power, war being an extreme example,[252] and seem to be the prototype of non-cooperative behaviour.

Cooperative behaviour is not immediately observed by actors as a prerequisite for achieving their goals.[253] Especially if more rounds are to fol-

---

[242] See Haley, 2002.
[243] See Maoz, 1989, p. 261.
[244] See Coddington, 1966, p. 522. Such situations are called tipping point or brinkmanship situations.
[245] See Cross, 1965, p. 71; Külp, 1965, p. 50.
[246] See Carr, 1968, pp. 143 ff.; Zartman, 1978.
[247] See Schelling, 1956, pp. 282 ff.
[248] See Yankelovich, 1982.
[249] See Deutsch/ Krauss, 1962, pp. 53 ff.
[250] See Hart, 1976, p. 294.
[251] See Boulding, 1963, pp. 427 ff.
[252] See Hess, 2003; Humphreys, 2002.
[253] See Lewicki et al., 1994, p. 38.

low, the superior strategy is a cooperative style.[254] Motivation, length of interaction, the strategies of the other actors, overt communications and personality characteristics are thus the major factors supporting cooperation. An institutional framework can help to back up cooperation, as can the anticipation of lower transaction cost.[255] 'Tit for tat'[256] is the principle used by actors who are trying to make the other actor move in a direction which will enable both actors to arrive at the best common result. It is a friendly mechanism, showing determination to cooperate: if the other actor is not willing to cooperate then the same amount of hostility in expended. Cooperation gives the actors the opportunity to focus on their core competence and hence even promotes innovation.

The interaction of actors 'equipped' with different strategies for either cooperating or not cooperating causes dynamics in negotiations.[257] Negotiations in general undergo a typical dynamic cycle, which repeats itself with certain variations.[258] Depending on what one actor decides to do, the reaction of the other will have a result which could have been drastically different if other strategies were to be applied by any of the actors involved. The attempt to predict the strategies which are most likely to cause conflict might lead into a vicious circle: 'what happens if I do this and he does that. He will then react this way, so that I have to react that way in order to anticipate his movement there.'[259] This is called 'the rule of anticipated reactions': the mixing up of past and future (expected) behaviour.[260] It is a 'loop function', where one reaction causes the next and so on.[261] Since there is no stable focal point except at the extremes, all of the possible intervening solutions might be experienced in the shifting back and forth between positions.[262] The actor that 'moves' first, might have a first mover advantage, though. On the one hand the struggle to avoid the resulting shifts is likely to lead to even more shifts, but on the other hand it could be that the different strategies of the different actors eventually cancel each other out in respect to power.[263] So what is the effect of strategy on power asymmetries in negotiations? The application of strategies causes shifts and asymmetries in negotiations. The uncertainty caused by the in-

---

[254] See Gibbons, 1992.
[255] See Voigt, 2002, p. 17.
[256] See Axelrod, 1988, p. 4.
[257] See Wagner, 1995, p. 11; Zartman, 1974, p. 398.
[258] See Erlei, 1998.
[259] See Schelling, 1957, p. 20.
[260] See Simon, 1957, p. 67.
[261] See Zartman/ Rubin, 2000, p. 284.
[262] See Schelling, 1958, pp. 208 ff.
[263] See Emerson, 1962, pp. 33 ff.

stability seems to be fairly evenly distributed between the actors. None of the actors knows exactly what the other actors are going to do and thus none can derive more profit from this situation than another because the next shift could result in an opposite distribution of power, like a pendulum which swings back and forth. The actors do not know where the pendulum is at any given time, because the speed at which it oscillates is unknown.

Summing up the issues of non-cooperation and cooperation shows that power connotes more than merely force and violence – it is on the contrary part of the common hierarchical systems in negotiations. Cooperation could be a strategy for limiting conflict and uncertainty. The likelihood of an agreement is higher when actors cooperate. Cooperation, however, first needs to be 'learned'. Before actors cooperate they try to make sure that no one is going to cheat, for example, by setting the right incentives to cooperate – a process which might be costly.[264] The distribution of power is not stable, but changes constantly as a result of actions and reactions. The power distribution in a negotiation at any one time represents only a snapshot which is not necessarily of help in future situations.[265] Asymmetry in this sense is not a stable state, but it is more stable than a symmetrical distribution of power. Asymmetries are the 'normal' condition, symmetry is the exception, and the latter only exists for short periods of time. This might be one reason why actors seek to distance themselves from the extremes. Few actors willingly risk all or nothing when determining their negotiation strategy. The risk factor would be too high. But still it seems as if actors are tossed on the horns of a paradox, where they have to decide to be either non-cooperative or cooperative.[266] It could make sense for actors to cooperate in some areas, but still to maintain a degree of competition in others: this is called '**coopetition**'.[267] With coopetition the high risks associated with using strategy without knowing exactly what the result will be, are limited. This kind of behaviour also respects the interdependencies between the actors, even if they are not visible to their whole extent.

Relative scale as an indicator. Besides its use as a tool, the second key idea of game theory is to deliver an indicator which supports an actor in a negotiation.[268] In general, models can work as an approach to describing the position of an actor in his negotiation: a simple method for finding out more about one's own position is prejudice. The individual negotiator cali-

---

[264] See Deutsch/ Krauss, 1962, p. 53.
[265] See Bartos, 1967, p. 493.
[266] See Hauser, 2002.
[267] See Nalebuff/ Brandenburger, 1996.
[268] See Allas/ Georgiades, 2001; Baumol, 1986, p. 9.

brates his own 'scale' with reference to a prototype, i.e. within his own frame of references.[269] It is the most widely used concept in everyday life and an indicator that helps actors cope with complex situations. Abstract models are more useful for solving the details, as they support the actor by splitting up a big problem into many smaller ones. This could help to identify dynamics in negotiations and might assist the actors in forecasting the results of their negotiation. But applying abstract models is a complex issue as well. In one way a model can help the actor to get to grips with the situation but it might also lead to irrational results, because only a fraction of the whole situation is taken into account. This could put the actor on the wrong track for his prognosis.

How is such an indicator set up? As the exact results of a negotiation cannot be predicted ex ante, an approximation of possible outcomes through an analysis of the extremes could be an alternative.[270] An actor will sketch for himself the best case vs. worst case scenarios. The best case would be that he achieves 100% of his goals, the worst case scenario is that the negotiation fails altogether. The actual result of the negotiation will probably be somewhere between the two extremes. Using this sketch as the basic structure, actors design their strategy: by trying to exclude, step-by-step, the worst possible results, actors try to get closer to the best possible result. The use of utility functions makes it possible to show all the possible points of agreement.[271] Game theory can help to describe such a 'contract zone',[272] making the instruments of power more transparent. In reality negotiators obviously settle for suboptimal results.[273] This could be due to the practical problem that such an indicator is needed for more or less each individual negotiation situation in a dynamic environment and cannot be generalized.[274] As a reaction to this practical limit actors tend to picture themselves in relation to the position of the other actor, instead of measuring absolute power ('I have 10 units of power and the other has only 5' might be their irrelevant conclusion).[275] In this respect power is a perceived relation ('I am more powerful than he is').[276] For the actors it is not important to know about the existence of power. They want to be able to make comparisons on the basis of different options, to see who is stronger in a

---

[269] See Kahneman, 2000, p. 707 ff.
[270] See Kahneman/ Knetsch/ Thaler, 1986, p. 299.
[271] See Nash, 1950.
[272] See Coddington, 1968, p. viii.
[273] See Young, 2001, p. 1.
[274] See Simon, 1966, p. 17.
[275] See Parsons, 1966, p. 79.
[276] See Zartman/ Rubin, 2000, p. 11.

certain situation.²⁷⁷ One limitation of this approach could be that when using a relative concept the actor may lose his overview of the situation: possible connections to other negotiations in the past or in the future are severed if only the currently available options are considered. In addition, a dynamic environment makes it difficult to evaluate a deal, especially for those who are not part of it. This might explain why the different actors often have different perceptions as to the value of a deal. Still, relative indicators support the actors, as 'economies of scale' in the production of such relative scales can be realised benefiting from experience.²⁷⁸ It would seem as if classic negotiation rules are more helpful than any theory,²⁷⁹ as experienced negotiators can predict the probable outcome without using models.²⁸⁰

Résumé: Resulting power in the strategy dimension. At a strategic level the dynamics of power cause asymmetries. This nourishes the doubt that game theory is of value in negotiations and there are suggestions that personal judgment may be more helpful,²⁸¹ since game theory works like a 'black box' model. It is true that there is a big difference between the relatively simplified world of game theory and the real world of negotiations.²⁸²

Acting on the perceived power caused by the strategic positioning of the actors, the actors try to change their situation and take various steps to target the other actors' positioning. A first such step is the decision whether or not to negotiate at all, but game theory also provides the tools for defining the style of the negotiation, i.e. whether to cooperate or not to cooperate. If actors were now only to observe the strict theory of games they could be doomed to follow a trend of predictions and even intensify dynamics.²⁸³ In order to use strategy efficiently in negotiations, it is thus necessary to include the environment. In order to further improve the applicability of the descriptive results of game theory²⁸⁴ a normative supplement is needed. A combination of a (game theoretical) model and the personal judgment of the actors seems to be best suited for this.²⁸⁵ This is why ex-

---

[277] See Dahl, 1957, p. 205; Parsons, 1966.
[278] See Druckman, 1971, pp. 530 ff.
[279] See Franz, 1968, pp. 242 ff.
[280] See Harsayni, 1956, p. 145.
[281] See Allas/ Georgiades, 2001, p. 87.
[282] See North, 1992, p. 18.
[283] See Young, 2001, p. 2.
[284] See Young, 2001, p. 45.
[285] See Allas/ Georgiades, 2001, p. 97.

perienced negotiators use the rational approach of game theory for some parts and employ other rather normative tools to fit these results together.

By showing the other actor which negotiating style is chosen, the instruments of power are made transparent, and can hence unfold their effect.[286] Once the instruments are transparent, **resulting power** gives the actors the opportunity of achieving their individual goals. A relative concept allows one to compare different available options. The actor can make his individual strategic decision based on alternatives. Even when strategy is transparent, the decision as to which individual option will be executed still lies with the actor. Thus the initially perceived limits of bounded rationality can be turned into an opportunity to design new alternatives of actions.[287] That is why economists use game theory to look at individual behaviour first and from this construct a theory of interaction between individuals and groups.[288] This also explains why most definitions of power are based on a relative concept and some even define the concept of power with power: this is not merely tautological, but suggests that a relative rather than an absolute concept is best suited to the description of power.

This leads to **Hypothesis 8**: *Game theory works as an indicator which enables one to grasp the dynamics of power in negotiations.*

Actors see a relief in an individual concept of cooperation and competition

Realising the source and making use of the dynamics of power, i.e. aligning the strategic powers, is what strategy in negotiations is all about. The problem is – and this is true for the framework and cost dimension as well – that the transparency of these actions is in most cases non-existent. This highlights the importance of information with respect to power and points to the difference between **perceived** and **resulting** power.

The power of strategy is the dynamic connection of all three power dimensions where the limits are set by access to 'information' and 'time'. Strategy is thus both a source of power and a cause of dynamics. The use of the strategy could be a sign that an actor is of the opinion that he is more powerful. It is not so much that one actor may have more or better access to strategy, but that one actor is more willing, ready, and able to use strategy. An actor who believes that he occupies the more powerful position could be more willing to make use of it. It is difficult to say whether or not

---

[286] See North, 1992, p. 21.
[287] See Simon, 1982, p. 419.
[288] See Buchanan, 1966, p. 25.

the strategies implemented by the more powerful actor are better or more efficient than those implemented by the other actor(s). **Real power** can be rendered transparent with the help of a relative, possibly a coopetitive, approach; and game theory could form the basis of such a concept. Applied correctly such a combined approach would leave room for individual decision-making based on a negotiation as to how to handle power as an instrument in the individual situation to the actors' mutual advantage.

This leads to **Hypothesis 9**: *Strategy, as commonly applied, leads to confusion and makes efficient negotiations difficult or impossible.*

The key characteristics of power in the strategy dimension that will be looked at in the case studies are the transparency of time and information as the source and cooperation/ non-cooperation as a cause of the dynamics of power.

2.3 Common characteristics define power in negotiations

The main focus so far has been on explaining how power affects negotiations and on describing the development from perceived to resulting to real power. The goal was to find a workable definition of power and negotiation to help identify the characteristics of negotiations that help not only the individual actor ('big' or 'small') but society as a whole to run negotiations in an efficient manner. Common characteristics of negotiations have emerged. Asymmetry is both the source for power and cause for dynamics.

This leads to **Hypothesis 10**: *The characteristics of negotiations and power are closely interrelated.*

The major sources of power seem to be the factors time and information. Dynamics of power seem to be caused mainly by the actions of the actors themselves. This is because actors adapt their power-behaviour, i.e. the application of power, to the power-behaviour of the other actors. It does not seem to be asymmetry that causes problems and which make negotiations inefficient, but the application of power as an instrument without regarding the interdependence of the actors.

The most important lesson to be learned from the positive analysis is thus that power is not a constant function.[289] Power can only be grasped by way of a dynamic model: chapter 2 showed that the different power dimensions are closely interwoven; it is even hard to describe them separately. Those who are trying to gain planning security in negotiations with

---

[289] See Zartman, 1974, p. 398.

the help of predictions are entering difficult territory. Since the Power-Matrix is thought to represent the source and the dynamics of a negotiation it needs to be discussed if there is only a single starting point. On the one hand it seems logical that the starting point is set with the 'framework' dimension, but in situations where the framework has remained unchanged for a while a starting point could be in the 'cost' dimension. In this respect it is important to consider the number of rounds of the negotiation: it is probably necessary to assume that there will be several rounds as one cannot exclude the possibility that after one round of negotiations – even with a result that benefits every one involved – the next round will commence. In any case, from the perspective of the actors there is likely to be more than one round. In the present study it is thus assumed that negotiations (in network) markets are **not** one-off negotiations. Most of the issues discussed in these markets are very complex and in the long run the actors usually have to depend on each other in order to reach a long-term and lasting solution. 'One-level problems' exist[290] but are not the focus of the present study. For the matrix this means a single passage through it is probably insufficient, and subsequent passages should not begin at the same starting point. For the purpose of explaining the theory and the structure of negotiations one passage should be enough, though.

The best concept to typify power for the purposes of this study is **defining power as the relative position of actors to each other in a given situation**,[291] since it is not clear precisely what a (strategic) investment in power causes. Since power is not absolute, only a relative concept or indicator can be used to measure it. The Power-Matrix delivers a structure to grasp power in negotiations for both the actor and the analyst.

---

[290] See Hauser, 2002, p. 39.
[291] See Dahl, 1957, p. 203; Hart, 1976, p. 293.

3 Negotiations in network markets

Case studies are needed to mirror the theory, but only a few parts of reality can be tested. In the following chapter the cases will be analysed with the help of the positive theory set up in chapter 2, so as to enable the drawing of normative conclusions in chapter 4. The goal is to obtain a broad understanding of the underlying mechanisms of negotiations in markets with asymmetrical power distribution. To be able to do so, both the international literature and actual (European) cases are evaluated in order to get and give an overview of negotiations in network markets. In order to be able to even 'read between the lines' when looking at the (technically, legally and economically) complex cases, interviews were conducted with a few of the actors.[1] Neither the interviews with the individuals nor the naming of specific companies in certain countries should belie the fact that here coherences on a macroeconomic level are being searched for. The seemingly individual details are inserted to authenticate the analysis and the resultant findings.

In the following, firstly, in order to gain a better understanding of the telecommunications and the rail transport markets, these will be described in more detail. Secondly, the Power-Matrix is 'applied' and tested on these two network markets by identifying both the sources and the dynamics of power and discussing the following questions: what role does regulation play and how can one gain access to the different levels of influence in a negotiation? Is this market characterised by a situation of 'big' vs. 'small' actors based on essential facilities where power is only an illusion? Which strategic instruments are available and what result do they have on the co-operation of the actors? Thirdly, the relevance of these finding for negotiations in other (network) markets will be discussed.

---

[1] See Werner, 2002; Neuhoff, 2003 and 2005; Hirschfeld, 2003; Rochlitz, 2003; Scheurle, 2003; Heinrichs, 2004; Wünschmann, 2005.

3.1 Network characteristics in telecoms and rail

Network markets are special. To allow for a focused approach there will follow a discussion of what is so special about network markets – here especially the telecoms and the rail transport sectors – and why they represent typical negotiation situations with an asymmetrical power distribution.[2] In both the telecommunications and the rail transport market negotiations take place in a regulated environment, where one large incumbent has a major market share and many smaller companies try to access a fraction of the rest of the market. The question posed in the following chapter will be whether power exists in these markets and whether it influences the negotiations in the three power-dimensions of framework, cost and strategy. Do negotiations take place because of or in spite of the existence of market failures? What role do asymmetries between the actors play in the course of negotiations? Altogether it will be necessary to show where 'power' can be found in those negotiation situations and what conclusions for the improvement of negotiations can be drawn from this.

In this chapter of the study the telecommunication and rail transport markets will first be analysed separately. An outline of the history and a classification of the actors to describe the framework will be followed by an overview of the cost and technical issues of each market. To enable one to judge the strategic instruments of the actors, the competitive situation will then be presented. Finally, the need for negotiations in network markets will be summarised.

3.1.1 Telecommunication – a dynamic market

What is special about the telecommunications market in respect to negotiations?

Framework – regulation to liberalise the telecommunications market. Historically, the main driving forces of change in the telecommunications market were technical innovations, namely the switch from analogue to digital equipment.[3] This in turn incited politicians to push ahead with liberalisation in the respective national markets.[4] National administrations had no choice but to implement the liberalisation – if (only) incremen-

---

[2] See WIK, 2005; Newbery, 2000, discusses also more network markets.
[3] See Kahn, 1988, p. 127.
[4] See Bundesministerium für Wirtschaft und Arbeit, 2003, for an international overview on the status of the liberalisation in the telecommunication markets.

tally.[5] The initial steps of liberalisation were taken during the period 1982-1989 in the US and the UK by splitting up the integrated telecommunication companies into the so-called 'baby bells' in the US (separating the incumbent operator into long-distance and local operators) and by setting up a duopoly (one incumbent and one 'new' player) in the UK. During the period 1989-1997 the telecommunications market in the whole of the EU was liberalised with the exception of public telephony. Progress was made in accordance with the relevant directives and green papers of the EU-commission.[6] Also during this time, when telecommunication in most (European) countries was still considered a part of the postal services, these sectors were formally separated. In Germany, for example, in 1989 the different entities were legally split up and the hardware market and the business customer services were liberalised, all on the basis of the so-called 'Poststrukturgesetz'.[7] Only in the third phase from 1998-2003 were the public voice telephony markets in the EU fully liberalised. National Regulatory Authorities (NRA) where established on the national level in order to implement and supervise the changes.[8] Most of the NRAs have 'mother- or sister-' institutions on the national level, such as cartel offices, because telecommunications regulation is sector specific, whereas the other institutions are general competition authorities. In Germany, for example, the 'Regulierungsbehörde für Telekommunikation und Post' (RegTP[9]) is the sector specific regulator for the telecommunications market and the duty of the cartel office is to supervise the market in general in regard to competitive questions.[10] The international authorities (in Europe the European Commission, globally the WTO[11]) monitor the liberalisation developments, respecting the sovereignty of the states. In the telecommunications sector negotiations are explicitly seen as the 'primary choice' and thus the obligation to negotiate, albeit only nominal, has been enshrined in

---

[5] This paragraph builds on Arnbak, 2005.
[6] See Heng, 2005, p. 3, for a complete listing of the relevant green-papers.
[7] See Büchner, 1990.
[8] See Schütz, 2005a, pp. 166 ff., for an overview of the key decisions of the German NRA.
[9] RegTP was renamed (and restructured) to become 'Bundesnetzagentur' (BNetzA) on July 13, 2005. In this thesis the abbreviation RegTP will nevertheless be applied, since all the decisions and actions described here were still made under the name RegTP.
[10] See Bundeskartellamt, 2002.
[11] See Fredebeul-Krein/ Freytag, 1999.

law.[12] Still the exegesis of the law leaves room for discussion on the actual negotiation level, where contracts are negotiated between the carriers and supervised by the NRAs. Telecommunications, more so than the postal sector, has since then developed from a purely monopolistic into a privatized, regulated, and highly competitive market.[13] The fourth phase, which is characterised mainly by the implementation of the so-called 'EU-99 Review' as a new framework demonstrates that in due course the international authorities gained more and more influence on the national level.[14] In Germany, for example, the most recent amending law ('Novelle') of the national telecommunications act ('TKG 2004'), revamping the German telecommunications act of 1996 ('TKG 1996'), was initiated by and based mainly on this EU-framework. It is expected that the next phase will start in 2006 with the 'review of the EU-99 review' where more than before emerging technologies, such as Voice-over-IP (VoIP)[15] and convergent products, such as the combination of different products (so-called Triple Play i.e. integrating voice, video and data services on a single broadband connection)[16] and the decline of conventional telephony (so-called 'POTS') will be the focus of the adaptations.[17]

Cost – networks as bottlenecks despite large number of actors. From an economic and technical point of view the telecommunications-market is a network market. The quality of the products (minutes of traffic) sold by the different operators and providers are almost identical; they can hardly be differentiated by the customer. This is true for the quality of the actual product (minutes) but not for the quality in terms of delivery times. Traffic on the networks needs to be synchronized in the dimension of time. That can be helped by (re-) routing in the dimension space, since the traffic is not bound to a specific route through the network. Obviously, the origina-

---

[12] See Commission of the European Communities, 2002b, Article 4 (1); Heun et al., 2002, p. 426, for the German telecommunications market; in the German TKG 2004 this is regulated in § 16.
[13] See Ritter, 2004; Büchner, 1990; Jäger, 1994, Knieps, 2004, for an overview of the developments, especially in the German telecommunications market.
[14] See Scherer, 2002. Tele2, 2005, offers an international overview of the regulatory situation in Europe. See Coen, 2005, for a critical view on the convergence into a single European regulatory model.
[15] See Schütz, 2005a, pp. 189 ff.; Kurth, 2005.
[16] See EITO, 2005, pp. 93 ff.; Arthur D. Little, 2005.
[17] See EITO, 2005, pp. 12 ff., for a description of the dynamic development of the telecommunication markets worldwide.

tion (source) and the termination (target) channel the traffic in the networks and make them the bottleneck of the whole transport system.[18]

The markets can be roughly divided into the fixed and the mobile network. The fixed networks can be subdivided into local and long distance networks. The firms in the telecommunications market can be classified into companies that operate their own networks or as those that run services only. Value added service providers can carry on their business 'on top' of all kinds of networks.[19] The setup of the associations in this market (VATM[20] is an example for Germany[21]) mirrors this (national) market structure: in the fixed network market there are many rather small actors that compete with but also depend on the networks that the incumbent (in Germany Deutsche Telekom, DT) owns and operates.[22] In the mobile market there are only a few companies that share each national market more or less equally between them.[23]

Strategy – competitiveness of new technologies. Competition in the telecommunications markets is fierce.[24] Potential strategies for circumventing the bottlenecks exist but have not been implemented in any substantial fashion. Substitutes for the bottleneck facilities of the infrastructure have been tested, but none of them has made a breakthrough possible (yet):[25] neither Cable TV[26], nor power-line[27] nor the wireless local-loop (WLL)[28]. On the product level possible substitutes for traffic via conventional net-

---

[18] See Bijl/ Peitz, 2005, for an international overview of the (regulatory-) problems of this bottleneck, mostly discussed under the heading 'local loop unbundling'.
[19] See Elixmann/ Schäfer/ Schwab, 2004, for an overview of the German value added services market.
[20] VATM originally stated in its constitution that the members are to work 'against' the incumbent Deutsche Telekom.
[21] See Putz & Partner, 2004, for a complete listing of the associations in the telecommunications market in Germany.
[22] See Dialog Consult/ VATM, 2004; see RegTP, 2005a, for data on the competitive market situation in Germany. See Scholl, 1998; Büchner, 1999, for the political background. See Neumann/ Stamm, 2002, for an overview of the discussion on further reform.
[23] See Koenig/ Vogelsang/ Winkler, 2005.
[24] See Hempell, 2005.
[25] See Monopolkommission, 2004a; Soltwedel et al., 1986, p. 198.
[26] See Rickert, 2004; FAZ, 2004d, p. 14.
[27] See Koch, 2004; Commission of the European Communities, 2005b.
[28] See RegTP, 2005a and 2003b; Monopolkommission, 2004a.

works are mobile communication, WLAN[29], Voice-over-IP (VoIP)[30], e-mail and SMS. Some of these products are having a noticeable impact, but not as a substantial substitution for conventional telephony.[31] On the one hand it is argued that the mobile sector is about to substitute the fixed-network market as the number of subscribers and the saturation of the market is in most countries now higher than in the fixed network markets.[32] On the other hand the mobile market delivers a different kind of service (most obviously there is to date no reliable, broadband internet access) and targets a different group of customers (always on, always direct line).[33] Other markets, such as postal mail are not a substantial substitute for telecommunications, but rather a complement to it. The main area of competition lies thus within a fairly small product dimension and is generated by only a limited number of operators both nationally and internationally.

All in all, the telecommunications sector is a very dynamic market, where negotiation powers are distributed asymmetrically.

3.1.2 Rail – a sedate market

What is special about the rail transport market in respect to negotiations?

Framework – regulation to free the market of its depression. Rail transport is considered a 'universal service',[34] which no private company would or even could supply.[35] For that reason the rail transport market was, and still is in most countries, owned and run by the state.[36] The driving force for change and the reforms in Europe[37] were the declining level of rail transport and the rising deficits that were about to get out of hand.[38] In

---

[29] See Frost&Sullivan, 2004, who state that WLAN could be a product which moves in between the two products fixed and mobile. This could hinder the substitution even more, as it drives the markets further apart. At present however, it seems as if it would connect the two markets.
[30] See Lunden, 2005; Kuthan, 2004; Sisalem, 2004.
[31] See RegTP, 2004b.
[32] See Golvin, 2004; Rodini/ Ward/ Woroch, 2003; Deutsche Bank, 2004.
[33] See Frost&Sullivan, 2004; Koenig/ Vogelsang/ Winkler, 2005.
[34] See OECD, 1998, pp. 191 ff.
[35] See Polatschek, 2004; Die Bahn, 2005a, allows for a technical overview of this complexity of running a railway.
[36] See Kirchner, 2003, for an overview of the European rail transport markets.
[37] See Friebel/ Ivaldi/ Vibes, 2004, p. 2; OECD, 1998, pp. 141 ff. See Miller, 2003, for a historic worldwide overview of the economic development of the rail transport sector.
[38] See OECD, 1998, p. 147.

Germany, for example, a study analysing the competitiveness of the state monopolist (at that time Deutsche Bundesbahn (for West-Germany) and Deutsche Reichsbahn (for East-Germany), now Deutsche Bahn, DB),[39] which was commissioned in 1989, came to this conclusion:[40] if DB were not be reformed soon, it would have no chance of ever getting rid of its debt.[41] Germany resembles the situation of most European countries and is therefore a qualified example. To counter the problems the first phase of the reform (in Germany the so-called 'Bahnreform' of 1994) was implemented, together with the introduction of the first law for the rail transport sector, the so-called 'Allgemeines Eisenbahn Gesetz', (AEG).[42] The main focus of the first 'Bahnreform' lay on the separation of the business units and the state supervised parts ('hoheitlich'). The Federal Railway Authority ('Eisenbahnbundesamt', EBA) was made responsible for handling the state tasks and the supervision of the 'Bahnreform'.[43] With the second 'Bahnreform' in 1999 the company DB was split up into separate enterprise units, but the infrastructure (rail tracks) and services were not separated. The goals of the 'Bahnreform' were to increase the relative proportion of rail transport traffic and to attempt to minimise the burden on the state budget.[44] The AEG obliges the incumbent to negotiate with his competitors, but leaves room for discussion on the actual negotiation level, where contracts are negotiated between the carriers and supervised by the EBA as a regulator.[45]

For the European traffic market it is the EU framework from 2001 – which should have been implemented by March 2003 in every single member nation – that sets the framework for today's rail transport market in Europe.[46] In Germany this new law was implemented as recently as in 2005, and includes the establishment of a new regulatory authority (starting operations in January 2006 as part of RegTP) overseeing the rail trans-

---

[39] See OECD, 1998, p. 12, for more details on this study called 'Untersuchung Wettbewerbsfähigkeit Bahn'.
[40] For the reform process of the rail sector in Germany see: OECD, 1998; Bennemann, 1994, Julitz, 1997, Berndt/ Kunz, 2000. For a critical view see VCD, 2004a; Ilgmann, 2003.
[41] See OECD, 1998, p. 12.
[42] See Bundesregierung, 2002a; PSPC, 2003; Bundesrat, 2005, for the reform of the AEG.
[43] See EBA, 2004; Schweinsberg, 2002.
[44] See Allianz pro Schiene, 2004, for a critical appraisal of the reform. For the political dimension of the German liberalisation see Friedrich, 2003.
[45] See Brauner/ Kühlwetter, 2002, p. 493; in the AEG this is regulated in § 14 (4); Schweinsberg, 2002.
[46] See Kirchner, 2003 and 2004, for an overview and structure.

port sector.[47] On the international level the EU passed in 2004 the so-called 'second European Rail-package' intended to additionally liberalise cross-border rail passenger service and improve quality in rail freight services by January 1, 2007.[48] In this context a European railway agency will be set up which is supposed to secure and support the technical standards necessary to run the interconnection between the networks more smoothly.[49]

Cost – networks serve the needs of incumbent and niche operators. Rail transport markets are considered 'special' in economic terms, because the network presents a bottleneck: many firms require access to it, but only one owns and runs it.[50] This is the reason why in Germany the rail incumbent DB (still) operates under an exemption rule of the cartel law.

The structure of the rail transport associations (in Germany, for example, the 'Verband deutscher Verkehrsunternehmen', VdV) gives a representative picture of the market participants: the number of public operators in Germany is 30 and the number of competitors is 293.[51] Even though this number seems relatively large, the market share of non-state-owned railways (NE) was only 4-5% of the traffic mileage in 2004.[52] The rail transport markets are thus characterised by the existence of one large incumbent and much smaller niche competitors.[53] More than telecoms the rail transport market is still a national market.

Strategy – intermodal traffic substitution. Actors in the rail transport market mostly compete and cooperate at the same time especially as the competition is not so much from other rail transport operators but rather from other kinds of traffic.[54] This intermodal traffic substitution away from the railways is constantly growing.[55] The reaction of the actors has been mainly to expand their traffic solution product portfolio to include other kinds of traffic, as the example of the incumbent DB, which took over the

---

[47] See Ilgmann, 2001, for the genesis of this new regulator; Rochlitz/ Winkler, 2005.
[48] See BDI, 2004.
[49] See FAZ, 2004b, p. 13.
[50] See PSPC, 2004, pp.43 ff.
[51] See Die Bahn, 2005b.
[52] See PSPC, 2004, pp. 20 ff.; Die Bahn, 2005b; Schweinsberg, 2002.
[53] See Grabitz, 2004; Machatschke, 2002, for an example of such a 'local' carrier:
[54] See OECD, 1998, pp. 249 ff.; Allianz pro Schiene, 2004, p. 2, for a discussion on the intermodal competition in the rail markets.
[55] See PSPC, 2004, pp. 10 ff.

logistics service provider Stinnes, demonstrates.[56] Smaller operators additionally develop niche offers instead of expanding rail transport in total.[57]

All in all, the rail transport market can be characterised as a sedate or even declining market, where negotiation powers are distributed asymmetrically.

3.1.3 The need for negotiations in network markets

Both the new, rather dynamic and also the old, rather sedate network markets present special characteristics which seem to be prone to asymmetrical negotiation situations. Even though these markets seem to be fundamentally different at first,[58] in respect to negotiations they seem to have a lot in common: the need for negotiations as such thus exists in these markets and is recognised by the actors,[59] but still negotiations do not always take place. It is the (extraordinary) need to negotiate due to the asymmetrical distribution of power which makes these network markets interesting for this study.

Both the telecommunication and the rail transport market developed over a long period of time, from monopolistic markets to (partly) liberalised but newly regulated markets. Looking at the markets today can only deliver a snapshot, but is a means to understanding their development over time.[60] The difference in 'speed' between the two markets points to the importance of the dimension of time and can even lead to a better understanding of negotiations in other markets.

3.2 Power in network market negotiations

Is it possible to recognize power not only theoretically but also in case studies? If it is possible, can a process from perceived to resulting to real power be identified?

In the following each of the three power dimensions is analysed especially in respect to the possible sources and dynamics of power. The goal is to find out if it is asymmetry that causes problems in negotiations.

---

[56] See Stinnes, 2003.
[57] See Die Bahn, 2005b.
[58] See Miller, 2003, p. 58.
[59] See Scherer, 2002, p. 283
[60] See Miller, 2003, p. 13.

3.2.1 Framework – the different levels of activity in negotiations

Regulation is an essential part of the network markets. This points to the key characteristics of negotiations in markets with an asymmetrical distribution of power between the actors in the framework dimension as they were analysed in the theoretical chapter: the property rights distribution as a source of power and incomplete information as a reason for the dynamics of power. This leads to the following questions: what is the role of the framework and of property rights in the network markets rail transport and telecommunications with respect to negotiations? Which levels do the actors activate to achieve their goals even in the absence of complete information?

The framework in network markets is called regulation and provides a source of power

Is power in the network markets distributed asymmetrically due to the property rights and is regulation a source of power?

Property rights, values and rules make negotiations possible but also cause them. Actors deem a framework in network markets to be necessary and see the need to respect it.[61] Such a framework builds upon property rights which in these network markets are constituted both on an international and national level; they are also seen as a source of power by the actors themselves.[62] The regulators, active on both levels, are the watchdogs of the given property rights distribution. Whilst, for example, the telecommunications sector is seen as a driving force for innovations of the whole society,[63] regulation to enable these innovations is requested by the market players.[64]

On the global/ international level property rights are set by the WTO[65] and on an European level the EU functions as a central regulatory power.[66] The degree of liberalisation varies from country to country and this makes it hard to compare them.[67] Obviously, and this is true for both network markets, the European framework is not adapted 1:1 by the member states

---

[61] See Heinrichs, 2004; Scheurle, 2003; Rochlitz, 2003.
[62] See Hirschfeld, 2003; Rochlitz, 2003.
[63] See SPD, 2004.
[64] See RegTP, 2004a; Tele2, 2005.
[65] See Fredebeul-Krein/ Freytag, 1999.
[66] See Kiessling/ Blondeel, 1998.
[67] See Kirchner, 2003 and 2004, for the railway markets.

but made to fit into the local markets.⁶⁸ A value based decision-making process going through the international/ European 'mill' of bureaucracy seems to affect and even change the original values, which in turn leads to a high degree of non-acceptance.⁶⁹ This demonstrates that on the international level network markets are characterised by regulation which is meant to "improve the incentives for network industries to perform optimally in terms of the (European) public interest."⁷⁰ To achieve this goal the international framework has established the primacy requirement for negotiations to resolve access, interconnection and dispute issues, for example in the EU directives, which are constantly reviewed and made transparent via the so-called implementation reports;⁷¹ this mirrors the concern of the framework about negotiations and market power.

The keywords subsidiarity and harmonisation describe the fact that the national regulation co-exists with the international regulation:⁷² national regulation interferes in markets at a more detailed level than does international regulation. More than on the international level, it seems, the sector-specific laws have an interface with the competition law⁷³ – again a sign of the laws being more detailed. A separate national framework is deemed necessary by the actors,⁷⁴ but discussion (now) revolves around the scope and duration of that regulation.⁷⁵ The overview below shows that the national framework level for Germany is very abstract, but the special laws are very detailed, leaving or even producing many 'grey areas'.

The national legislative framework in the rail transport and the telecommunications sector for Germany

The so-called coalition agreement of the German government ('Koalitionsvereinbarung') from 2002⁷⁶ sets the guidelines for the regulation of the **rail transport** sector as follows:

- The responsibility for the tracks will stay in public hands.

---

⁶⁸ See Heinrichs, 2004; Scheurle, 2003.
⁶⁹ See Hirschfeld, 2003; Scheurle, 2003.
⁷⁰ Pelkmans, 2001, p. 437.
⁷¹ See Commission of the European Communities, 2004d.
⁷² See ERG, 2003b, for an example of the working groups of the national regulators and their discussion of harmonisation of remedies to counter the market failures.
⁷³ See Ruhle, 2005, p. 11.
⁷⁴ See Yan, 2001.
⁷⁵ See Heinrichs, 2003; Scheurle, 2003; Rochlitz, 2003; Hirschfeld, 2003.
⁷⁶ See Bundesregierung, 2002b, pp. 16 and 44 ff.

- Non-discrimination will characterise access to the infrastructure.
- Establishment of a new regulator.
- Liberalised access to freight and public transport networks in the EU.
- Minimise bureaucracy of the legal situation in the rail transport sector.
- Secure quality standards.

On the **telecommunications** sector the following guidelines have been set:
- Continue to run a competitive telecommunications policy.
- Support the fast introduction of UMTS.
- Let the customers profit from low tariffs and a wide product portfolio.

This list of rather political issues shows that national 'starter packages' which influence the individual negotiation situation exist in both network markets. The presence of such starter packages does not yet make negotiations in these two network markets easier, but at least provides the conditions for their existence. In the German telecoms law not only the need for negotiations but also the aspects of arbitration and mediation are explicitly mentioned.[77] This demonstrates that regulators consider negotiations in these markets to be necessary, but at the same time even they point to the fact that negotiations will not be successful in every case.[78]

Rules of the game help to identify the different levels and sources of power. The property rights distribution assigns the actors positions in their markets.[79] Asymmetrical distribution of power between the actors is seen as a problem by the regulators,[80] as it obliges the actors to move away from the position they are in.[81] By trying to solve this problem regulators set the rules of the game. Consistency could be an 'obvious sign' as Schelling calls the rules of the game for the actors. In reality, though, different market segments are regulated and treated by the regulators in different ways. Network markets are more often than not asymmetrically regulated,[82] in an attempt to provide a level playing field, i.e. establishing uniform access to the different levels of a negotiation. Asymmetrical regulation does not merely restrict actors with more power, though, but can also support them.

---

[77] See Winkler, 2005; in the German TKG 2004 this arbitration is regulated in § 51 and mediation in § 124.
[78] See Kurth, 2002; Schütz, 2005a, p. 159.
[79] See Wünschmann, 2005; Rochlitz, 2003; Scheurle, 2003.
[80] See Kurth, 2002.
[81] See Bijl /Peitz, 2002, p. xi, for regulation in telecommunication markets. For a critical view on asymmetrical regulation in the broadband market see Crandall/ Sidak/ Singer, 2002.
[82] See Ritter, 2004. pp. 110 ff.

For example price-capping regulation dictates a regular price decrease but also guarantees the regulated company a certain price level for a certain time;[83] smaller actors might experience more difficulties than larger actors in obtaining such assistance.[84] Additionally the passing of time alone seems to favour 'big' actors, i.e. the incumbents, as 'small' actors are being slowed down to a (relatively) greater extent by new obligations for all actors.[85] For example, the duration of regulatory proceedings is seen as such an obstacle in the development of a free market in the telecommunications sector.[86] Regulation hampers 'big' actors insofar as they are obliged to obey more rulings and cannot move as freely as the 'small' actors.[87]

It is difficult to maintain consistent (pricing) regulation in a dynamic market, though.[88] For example, the mobile-carriers have so far almost totally avoided regulatory action.[89] The case of regulating mobile markets or the so-called mobile termination calls (those calls that originate in a fixed network and have to be handed over to a mobile network as the called party can only be reached there) shows that regulators all over the world react differently: in some countries, such as the UK and Austria, regulators have declared each and every mobile market to be a monopolistic market; in Germany the regulator pushed the actors to negotiate more agreeable rates in order to avert price regulation.[90] The example of the introduction of the new technology VoIP might illustrate the difficulties of consistency as well: VoIP challenges the basic and well accepted pricing principle 'calling party pays' in the European telecommunications market;[91] overturning it would have far-reaching implications for the whole market system. Especially in the transition period from one system to another there is then no way of keeping up the desired consistency. One method of creating consistency is to make use of benchmarks. To do so institutions in the network

---

[83] See Ernst, 2001.
[84] See Rochlitz, 2003.
[85] See Neuhoff, 2003.
[86] See Mai, 2004, p. 4.
[87] See Hirschfeld, 2003.
[88] See Nett/ Neumann/ Vogelsang, 2004; Freytag/ Winkler, 2003; Nolte/ König, 2005.
[89] See Wirtschaftsrat, 2004, pp. 2 ff., demonstrating that the mobile operators express a belief that it should stay that way.
[90] See Koenig/ Vogelsang/ Winkler, 2005; Crandall/ Sidak, 2004; Deutsche Bank, 2004; Rodini/ Ward/ Woroch, 2003; Frost&Sullivan, 2004; Wengler/ Schäfer, 2003; Monopolkommission, 2004a, for the fixed to mobile termination issue and the regulatory problems around it
[91] See Winkelhage, 2005, p. 20.

markets constantly analyse the market, for example with the help of scorecards. The 'ecta regulatory scorecard'[92], for example, compares the general powers of the NRA, efficiency of the dispute settlement body, application of access regulations, the availability of key access products and the extent to which the New Regulatory EU-Framework has been implemented. It shows that there are big differences between the different countries and institutions in how the results of the negotiations are valued by the market participants.[93]

Regulation and its institutions deliver and guard a framework which is necessary for negotiations and is accepted by the actors as an anchor, even in a dynamic environment.[94] However, as different market segments are regulated to a different extent,[95] neither property rights nor regulation in network markets support the actors symmetrically or create a symmetrical playing field. In this sense regulation creates asymmetries at the same time as it tries to provide the actors with access to all levels of a negotiation and to establish 'obvious signs' which enable them to negotiate in the first place. But if regulation is asymmetrical, it is obvious that not all actors will accept the regulation because it will limit some more than others.[96] The incumbents, for example, see asymmetrical regulation in general as a danger to their activities and express this openly and defend their position heavily. DB considers the risk of regulation (what is meant are 'wrong decisions') worth € 2.5 billion p.a.[97]

Regulation creates institutions that make the market more complex. For both markets regulation is and has been a cornerstone of its development.[98] Since the markets evolve fast some segments are already competitive and new segments are created on a regular basis where it is uncertain whether regulation could support the market development.[99] Regulation creates institutions which try to mimic the complexity of the market; in this sense regulation has to follow the market and at the same time form it. The power balance between the institutions is not symmetrical. This is demonstrated by the fact that both network markets are regulated by national regulators with national peculiarities[100] but are under the supervision

---

[92] See ecta, 2004.
[93] See Ludewig, 2005, p. 19.
[94] See Rochlitz, 2003; Hirschfeld, 2003; Scheurle, 2003.
[95] See Giovannetti, 2005, p. 46.
[96] See Hirschfeld, 2003; Rochlitz, 2003.
[97] See McKinsey, 2002, p. 2.
[98] See OECD, 1998, p. 23 and pp. 169 ff., on regulation in railway markets.
[99] See Canoy/ Bijl/ Kemp, 2004, pp. 161 ff.
[100] See Engel, 2002a.

of the EU-commission.[101] The actors see for the European railway markets the national level as the most important level of influence.[102]

The complexity of negotiation situations increases with regulation. The following case of how a provider in Germany can possibly offer a rail transport service exemplifies this.[103] Firstly, the provider can apply for the right to offer a service where no subsidies are needed, i.e. a service which is profitable on the basis of the ticket prices. In this case the law provides ordinances that need to be obeyed in order to be able to operate such services. To obtain access to the tracks, negotiations with the incumbent or the company that owns the tracks are necessary. The kind of services negotiated this way are mostly long distance rail connections. Secondly, and this is the case for most of the local and regional traffic, the providers can apply for a tender-competition ('Ausschreibungswettbewerb'[104]) for traffic that does not support itself financially and thus needs subsidies paid by those who order the traffic service: the states ('Länder'), since in the course of the German Bahnreform the local passenger services were regionalized.[105] In this situation the rail transport market providers are confronted with at least three actors with whom they need to negotiate. These are the customers, i.e. the states, the incumbent and other new providers. The responsibility for financing the services lies with the states, which are refunded from federal funds. This creates a strong (financial) dependency between these actors in the traffic market, as the one that orders the services also pays subsidies for it.[106]

Even though the complexity of negotiations in general increases with regulation, regulatory reforms have had a positive effect on the efficiency of network markets.[107] It seems as if efficiency is improved when the regulatory measures are implemented sequentially.[108] Self-regulatory institutions, such as a standard reference interconnect offer (RIO)[109] in the telecommunications market, make negotiations if not obsolete certainly more

---

[101] See Kiessling/ Blondeel, 1998, p. 571.
[102] See Héritier, 2001b, p. 24.
[103] This paragraph builds on Schaaffkamp, 2004; Balzuweit/ Brümmer, 2002; Wiesmann/ Steinbach, 2002.
[104] See Rochlitz, 2002, p. 1, for a critical view on tender-competition.
[105] See OECD, 1998, p. 12.
[106] See Werner, 2002.
[107] See Friebel/ Ivaldi/ Vibes, 2004, p. 2, for the rail markets and Ritter, 2004, pp. 272 ff., for the telecommunications market.
[108] See Friebel/ Ivaldi/ Vibes, 2004, p. 4.
[109] See Ypsilanti/ Xavier, 1998, p. 653; in the German TKG 2004 this is regulated in § 23.

standardized.[110] The standard negotiation processes can in this way be further improved,[111] but can only be implemented when a certain stage of maturity is reached.[112] It is questionable, nevertheless, whether the state should virtually take over parts of the negotiations (by setting the RIO) instead of letting the actors themselves negotiate.[113] There is a general tendency in democracies to be seen to rule too much ('Verrechtlichung'). Some argue that this is the reason why actors might not feel they are being treated in a fair manner,[114] especially as the regulatory process is not transparent enough. Regulatory hearings have been one instrument implemented in the network markets in an attempt to make regulation fair by respecting the values of the actors and making the negotiation process more transparent.[115]

Résumé: Regulation is a countermeasure to perceived power. The framework for network markets has been identified as a source and at the same time as a cure for market failures. The actor who has the 'right' starters package, which is manifested in the network markets as regulation, is perceived to be in a better position to 'win' a negotiation.[116] In confirmation of **Hypothesis 1**, it can be stated that property rights are distributed asymmetrically between the actors and as such are a source of **perceived** power in negotiations. Incumbents, for example, are perceived to be 'on top', as property rights are not distributed symmetrically between the actors. But this is only an indicator for the amount of power that actors have, and not a measure for the success or failure of the use of power. If power between the actors is initially distributed asymmetrically, institutions try to level this out with the help of regulation. Regulation is thus a countermeasure to perceived power in network markets. But as regulation makes negotiations even more complex, this countermeasure can cause even more asymmetries. To summarise it cannot be stated what effect regulation ultimately has on the power distribution between the actors.

---

[110] See Stoffels, 2004, p. 6, discusses the possibilities of the operators to negotiate standard contracts and adapt them to their own needs with the incumbent DT. He finds that even though negotiations take place, DT is not willing to change core issues of the contract.
[111] See Better Regulation Task Force, 2001, p. 5; OECD, 1998, pp. 211 ff.
[112] See Ritter, 2004, pp. 273 ff.
[113] See Perrucci/ Cimatoribus, 1997, p. 508.
[114] See Susskind/ McKearnan/ Thomas-Larmer, 1999.
[115] See Scheurle, 2002.
[116] See Rochlitz, 2003; Heinrichs, 2004.

The need to change the framework causes dynamics of power

Regulation is a permanent challenge in the network markets and leads to shifts in the market positions of the actors.[117] Actors react to regulation by getting their negotiations going.[118] This leads to dynamics. In Germany alone in the last ten years two telecommunication laws, two rail transport laws and numerous laws and resolutions on the EU level were passed and implemented.[119] Additionally the regulator and the courts handled thousands of cases and passed rulings thereon.[120]

Actors cause dynamics, while their ability to process information is limited. Just as the actors are not all of the same size, not all wish to have the same kind of framework. Obvious win-lose situations in negotiations caused by an unequal distribution of property rights seem to provoke challenges to the framework by the actors.[121] Actors typically call for regulation of the other market players, but not for themselves.[122] Only where it is impossible to request regulation of the other players (for example, because of one's being the incumbent), is the call for less regulation altogether.[123] It is typically the case that if actors feel they are being treated unfairly or that they can easily get better results, they start to try to change the rules per se.[124] As rulings in the telecommunications sector by the regulator in Germany are commonly unacceptable to either side, often both actors appeal against them.[125] In what seems to be a rather cynical assessment, some note that if both sides complain the regulator must have made the right decision.[126] This only hints at the fact that the actors see such rulings as a way of creating not a win-lose situation but (even worse) a lose-lose situation. The most prominent example of such a (German) decision is the ruling on the new 'Element-Based-Costing' (EBC) regime by RegTP in 2002.[127]

---

[117] See Frühbrodt, 2002b, p. 14.
[118] See Scheurle, 2003.
[119] See Kirchner, 2003 and 2004, for an overview of the legal situation over time in the rail markets in Europe.
[120] See Handelsblatt, 2001, which shows that in 2001 there were about 1,200 cases in Germany where RegTPs decisions were appealed by operators. The exact numbers are not published.
[121] See Scheurle, 2003.
[122] See Hirschfeld, 2003; Heinrichs, 2004; Rochlitz, 2003.
[123] See Die Bahn, 2002b-2005b.
[124] See Rochlitz, 2003; Heinrichs, 2004.
[125] See Scheurle, 2003; Hirschfeld, 2003.
[126] See Scheurle, 2003.
[127] See Ritter, 2004, pp. 177 ff., for a detailed description on the genesis of the EBC ruling.

Why does a regulator make decisions that are not accepted by either side? It is most likely that regulators have insufficient information on which to base a decision.[128] Even though actors know that regulation is never complete and watertight as the regulator is not fully informed,[129] they still appeal to the regulator in the hope of a decision which favours (only) them. In this respect actors seem to be limited in their processing abilities because they would otherwise have proceeded to try to negotiate a solution which suited both actors better. This kind of reaction is like hiding behind a veil of ignorance:[130] actors reduce the complexity of their negotiation situation by outsourcing the decision to a regulator even if the regulator is going to base his decision on incomplete information and on a static model only.[131] In addition, actors also seem to behave irrationally: studies in the telecommunications market have shown that people who have children are the ones who protest most volubly against the erection of new masts for cellular phones; at the same time people with kids are the most frequent owners of cellular phones.[132] While on the one hand this is a good example of the irrational behaviour of actors, on the other hand it reflects the bounded rationality of the actors which from the outside are perceived to be irrational.[133] The term *Wohlbegründetheit*, introduced in the theoretical chapter, encapsulates the fact that the actors rely on the experience of others and in the network markets actors rely on the decisions of the regulator.

Together, these facts show that actors cause dynamics in negotiation situations not despite their limited ability to process information, but precisely because their ability is so limited.

Public choice helps to explain the existence of the many actors and why they use political market mechanisms. Actors make use of the different (political) levels for their cause.[134] The cost of understanding the framework and finding the right level seems to be high: in network markets the legal fees account for as much as 50% of the transaction costs.[135] Much of that amount is used to find out what the framework allows for, as

---

[128] See Gifford, 2003, pp. 477 ff. He even claims for the public utilities regulators that "it is a wonder that they can do their job at all."
[129] See Kiessling/ Blondeel, 1998.
[130] See Rawls, 1971, pp. 136 ff.
[131] See Canoy/ Bijl/ Kemp, 2004, p. 162.
[132] See Büllingen/ Hillebrand/ Wörter, 2002, pp. 65 ff.
[133] See Heinrichs, 2004.
[134] See Krummheuer, 2005b, p. 17, for the example of Connex, which publicly claims that they will go to court and to the EU commission to gain access to the network of DB.
[135] See Wünschmann, 2005; Scheurle, 2003.

3.2 Power in network market negotiations 83

not even the regulator knows how best to approach certain (regulatory-) situations in the markets.[136] This is due to the complexity of political processes: actors have no way of finding out how exactly their resources are transformed into output in politics. This is due to the number of (political) stakeholders who might bring influence to bear on one single negotiation situation. For both markets and their respective frameworks it is true that even outsiders that are not part of the actual negotiation (institutions, other competitors, bureaucrats, etc.) have the opportunity and an interest in influencing the negotiations.[137] For the telecommunications markets it has been shown that, for example, the capital markets are a major external driving force for the behaviour of the actors.[138]

The adaptations of the network laws in Germany are a good example of the long period of time needed to implement changes: the political discussion on and the acclamation of the laws started in the year 2001. In June 2004 the telecommunications law was finally implemented,[139] the rail transport law in March 2005.[140] These years were full of discussions on the possible details of such a law,[141] how much of the EU-framework needed to be adapted and especially as to whether regulation is or would be necessary at all. If it were necessary, how far should it go, how long should it last, and by whom should it be enacted? One of the main results is that the laws, by trying to cover more issues than before, became much more extensive, trying to leave fewer 'grey areas'.[142] But trying to make laws more explicit carries the danger of making them even more difficult to apply.[143] That the new German telecoms law is more detailed can be seen in the sheer number of paragraphs before the reform (100 in TKG 1996) and after (152 in TKG 2004), which could indicate that the longer the market is liberated, the more detailed the regulation will be.

In trying to influence the lawmaking process it was necessary for the incumbent Deutsche Telekom to intervene on each of the different levels: the national level, the EU level, and at the global level (via, for example, the WTO). Trying to get issues through brings enormous expenditures with it,

---

[136] See Heinrichs, 2004.
[137] See Neuhoff, 2003; Heinrichs, 2004; Scheurle, 2003.
[138] See Elixmann/ Metzler/ Schäfer, 2004.
[139] See Monopolkommission, 2004b. For a critical view see Heinacher, 2004,
[140] See Rochlitz/ Winkler, 2005.
[141] See Bundesrat, 2004, for an illustration on just how deep into details the discussion goes.
[142] See Cave, 2004, p. 27.
[143] See Ruhle/ Lichtenberger/ Kittl, 2005, p. 127, for Austria; Scheurle, 2003, for Germany.

as there are high institutional barriers of change.¹⁴⁴ For example Deutsche Telekom has offices on all the different levels to coordinate the lobbying process. The lawmaking process for the incumbent Deutsche Bahn is 'only' a European affair, since the (main) business takes place in Europe.¹⁴⁵ The existence of regional offices all over Germany could hint at the fact that rail transport is a regional business. This demonstrates that there are different levels of influence which need to be accounted for when trying to influence negotiations and the speed with which a measure is adapted into a national or international framework depends on the 'time phase' that the market is in.¹⁴⁶

The (economic) need to adapt the framework causes a political discussion which sets off dynamics in the negotiations of political actors which in turn influence the negotiations of the actors.¹⁴⁷ On the one hand there is thus a high risk that the political level makes regulation (and therefore negotiations) more complicated than necessary. On the other hand the construction of a framework, based on the political will, can help to establish a compromise between the politics and the economic necessities of a market sector. The theoretical concept of public choice could ensure the transparency of those points where influence is accessed,¹⁴⁸ by delivering an indicator which helps one to better understand the mode of action of the political process.

Capture establishes regulation as an end in itself, whereas politics and bureaucracy create dynamics in respect to time and content. Even though regulatory institutions are supposed to be independent, a certain political influence cannot be neglected.¹⁴⁹ This has an effect on the content and the time dimension of negotiations. The example of Germany demonstrates this: the incumbent DT is still partly owned by the German state, which is at the same time the political head of the regulator (here the Ministry of Economics). The political influence is hence built into the system.¹⁵⁰ The incumbent DB is still 100% owned by the State and the regulator is closely connected to the State (here the Ministry of Transport and the

---

[144] See Eberlein, 2001, p. 381.
[145] See McKinsey, 2002, pp. 2 ff., highlighting the importance of lobbying within Deutsche Bahn.
[146] See Commission of the European Communities, 2003, describing the time-phasing of monopolies.
[147] See Scheurle, 2003; Hirschfeld, 2003.
[148] See Neuhoff, 2003.
[149] See Neuscheler, 2005, p. 45; Jamison, 2005.
[150] See Wirtschaftsrat, 2003, p.3.

Ministry of Economics).[151] Both incumbents use their close political 'connections' to their advantage in negotiations on all levels: firstly, Deutsche Bahn, as an old state bureaucracy, has better contacts than any of its competitors and therefore enjoys an advantage regarding access to the existing contacts. DB invests heavily in local politics in order to maintain the contact with those who are responsible for awarding contracts on the local level.[152] These activities have even given rise to charges of corruption.[153] Secondly, DB has an enormous media-power by virtue of the (very) fact of its being the incumbent. Thirdly, it has close contacts with the scientific community and finally it has good contacts with the judiciary. Having access to all these different levels provides the incumbent with power, but also creates political dynamics. This can be exemplified by the political discussions surrounding the so-called 'Task Force' appointed to discuss the future regulation of the rail transport market.[154] Originally it was planned to set up a new and separate regulator to oversee the German rail transport market. This would have been a track-agency ('Trassenagentur') based on or within the structure of the EBA as a 'sister-office' to the cartel-office, with a very limited, but specialised scope of duties. During the (political) discussions the positions of the actors changed constantly.[155] In due course the actors seem to have lost contact with the basis of the original discussion and instead concentrated simply on consolidating their political power positions.[156] The danger is that the original impetus for change, after a political discussion has been started, no longer seems to be the main focus of the negotiation.[157] In this concrete example the discussion was finally detached from the question of the if, but only focused on which other regulatory body the new regulator should be attached to rather than what functions it was to have.[158] This demonstrates that politics were heavily influenced by the incumbent who was to be subject of such regulation.[159] For DB this was a very successful move, making use of all possible levels of influence, and averting the implementation of a new, specialised regulator. Such a situation is called a 'captured network'.[160] On the whole

---

[151] See Neuhoff, 2003.
[152] See Germis, 2004, p. 45.
[153] See FAZ, 2004a, pp. 13-15.
[154] See Keuter/ Waitschies, 2004.
[155] See FAZ, 2005c, p. 13; FAZ, 2005a, p. 11.
[156] See Neuhoff, 2005; Handelsblatt, 2005c, p. 4.
[157] See Rochlitz/ Winkler, 2005.
[158] See Handelsblatt, 2005b, p. 4.
[159] See Ilgmann, 2001.
[160] See Tomik, 2002, pp. 41 ff.

Deutsche Bahn seems to be swiftly balancing on the 'weak' factors.[161] It can also be stated that the political dimensions in both network markets in Germany, as in this respect they function in the same way, are noteworthy and have a significant impact on the negotiations in these markets.[162] 'Connectivity' arouses suspicions among those competitors who do not have such a network of contacts and who fear that it lessens their own chances in negotiations – this, in turn, causes even more dynamics.[163]

Besides the possible deviations from the original content-goals political influence prolongs negotiations: regulatory decisions take their time and are often delayed because of pending court decisions.[164] This is true for the German telecommunications sector where the telecommunications law explicitly orders the regulator to implement decisions, even if court cases are pending.[165] It is bureaucracy that puts the brake on negotiations as the following example shows: in Germany the question as to whether all traffic contracts in the rail transport sector have to be awarded on the basis of a tender-competition is by now the source of a high-level discussion fuelled by a series of contradictory decisions in the national courts.[166] According to the European ordinances on tenders the state is obliged to tender services that are worth more than € 200,000. This is to apply to almost any rail transport tender. Still many German states award long term contracts to providers, mostly the incumbents, without a tender.[167] To justify this the political term 'marktorientierte Direktvergabe' (market oriented direct tender) was invented in the course of the discussions surrounding this issue.[168] It is likely that this point will only be resolved in forthcoming court cases.[169] Only in the long run it is expected that it will become a definite rule that all such services will have to be awarded through a tender. That this (legal) process is time consuming can be seen from the timeline of the court cases (first court order in 1998, second court order of the European Court in 2003).[170] Together both the cases from the rail transport and the telecommunications market show that a framework implementation is a rather long term process. One argument for a slow transition is that the ac-

---

[161] See PSPC, 2004, pp. 51 ff.
[162] See Héritier, 2001b, pp. 7 ff.
[163] See Rochlitz, 2003; Werner, 2002.
[164] See Gifford, 2003, pp. 471 ff.
[165] See VATM, 2002.
[166] See Die Bahn, 2002b, p. 29. DB denies that long term contracts are signed in order to cut out the competition, but only to be able to make investments.
[167] See FAZ, 2002a, p. 15.
[168] See Schaaffkamp, 2004.
[169] See Werner et al., 2004.
[170] See Hirschfeld, 2003.

tors need time to change and to adapt to the new regulation.[171] In this respect it can be stated that transitions take time and are, if regulated, often very bureaucratic, which causes even more time lags.[172] What is even worse is that not only does a transition take time, but the original goals are not necessarily achieved.[173]

As the network markets have developed dynamically in the dimensions of time and of content the frameworks of the telecommunications and traffic markets have gone through the processes of liberalisation, privatization and regulation. This dynamic development of the network markets has also changed the need to negotiate.[174] There even seems to exist a built-in dynamic in regulated markets, moving back and forth between regulation and deregulation.[175] This is sometimes perceived as a paradox insofar as monopoly markets are liberalised but at the same time regulated, to be re-regulated and de-regulated over time.[176] The case study of framework issues above showed that the actors use the framework to 'develop' and protect their own situation once they have established their status. Regulatory regimes tend to develop their own life, as the market reacts immediately to changes and exemptions. An example in the telecommunications market of such a windfall profit situation on the basis of an exemption is the high price of local calls charged by the incumbent in Germany, as this was maintained as a regulatory monopoly until 2003. These cases also show that the actors can easily get the impression that the regulator is more interested in his own commercial issues than in developing the market. There have been cases, for example, where a regulator has almost invited actors to go to court against him.[177] That way a different (higher) level is also involved and the regulator can make a decision which cannot be blamed on him, but on the higher level. It is a public choice strategy to use different levels as a lever against other actors. The deadweight losses caused in this process are due to capture.[178] Owing to the existence of these capture effects it is often drastically concluded that reforms do not have the desired effect at all.[179]

---

[171] See PSPC, 2004, pp. 16 ff.
[172] See Scheerer/ Nonnast, 2005, p. 3.
[173] See Handelsblatt, 2005a, p. 11.
[174] See Lyon/ Huang, 2002, p. 112, point to the following order of actions in a regulated market: negotiation, conflict, regulation, then again negotiation.
[175] See Pelkmans, 2001, p. 436, for an overview of the dynamic development of the EU liberalisation.
[176] See Wegmann, 2001, pp. 50 ff., for this paradox
[177] See Scheurle, 2003.
[178] See Newbery, 2000, p. 141.
[179] See OECD, 1998, p. 22.

Résumé: By making use of different levels actors limit but at the same time cause dynamics. The negotiation situations in the two network markets show that there is a major trend towards non acceptance of the existing rules and framework. For this reason the actors try to turn the existing asymmetries to their advantage with even more activity. One problem is the time lag, as the speed with which the actors can react is, obviously, faster than that of the framework of the relevant market.[180] To pass this obstacle actors act on different levels at the same time. Thus, those actors who have access to more levels can exercise more long term influence over the framework of the negotiation situation.[181] It is the art of bringing influence to bear on the different levels, 'connecting' them where necessary, which allows actors to make the best of a particular negotiation situation.[182] The actors hence create a better basis for their negotiations.

To sum up the issues on the dynamics of power in the framework dimension, one can state that the actors both try to limit and to make use of the dynamics, but are constantly 'hampered' by incomplete information and bounded rationality. Asymmetry is a motor of this dynamic and regulation could provide a way out of this situation but could – just as well – cause even more dynamics. In confirmation of **Hypothesis 2**, it can be stated these dynamics cause **resulting power** which is then manifested by the different actors at each level of the negotiation.

The framework is a driving force to negotiate

It is the initial property rights 'starter package' which predetermines whether negotiations are at all possible in network markets. In the two network markets described here this initial setting is marked by a regulatory regime. The key characteristic regulation causes power. The cases show that the actors use the framework in order to identify the amount of power that is available to them for a particular negotiation situation. The framework in this respect functions as the source of power for the actor. The actors derive **perceived** power from the framework.

Actors are aware that the instruments on the different levels offer a lever for negotiations.[183] In trying to turn regulation to their advantage by reacting to it, actors find that achieving the right level of influence can help to 'level out' the asymmetry. These reactions come forcefully and quickly, irrational as it might seem at times. The actors would rather react than do

---

[180] See Ermert, 2004.
[181] See Werner, 2002.
[182] See Wünschmann, 2005.
[183] See Héritier, 2001a, p. 28.

nothing, they would rather move against each other with no predefined solution in mind than work together. It is not the asymmetrical distribution of power that lets the actors react in this way, but rather the uncertainty they face. The ability to move on to another level of the negotiation might be an efficient use of the resources for dynamic situations such as negotiations. Bounded rationality and incomplete information are the key characteristics of dynamics in negotiations on the framework level, and because of these the actor will never be sure that he can achieve the desired results: **resulting** power.

Where is **real** power in these cases? Only if the source of power and the dynamics are considered together, are the actors able to identify 'real power' in their negotiation situation. Those actors who understand how to interconnect the different levels that exist in the network markets can clear the way of obstacles so that both actors can profit from it. In response to **Hypothesis 3** this means that real power in the framework dimension allows the actors to use the different levels of a negotiation as an instrument of power. But very few actors seem to make (good) use of the different levels, as many negotiations fail to produce the desired results. To summarise, it can be said that negotiations take place despite, or because of, the fact that power is distributed asymmetrically between the actors. The framework for negotiations in these network markets seems to be one of the major driving forces behind the actor's decision to negotiate.

3.2.2 Cost – Profit from differences in size

Utilities such as telecommunications and rail transport both need a network to deliver their services. In both markets it is commonly the incumbent player that owns the majority of these networks. This implies that the different firms in each market segment have different levels of access to this scarce resource, but all depend on their ability to use it. In this respect both the telecommunications market and the rail transport market show characteristics of an asymmetrical power distribution.[184] The key characteristics of negotiations in markets with an asymmetrical distribution of power between the actors in the cost dimension were in the theoretical chapter shown to be transaction costs and essential facilities as sources of power and uncertainty as a reason for dynamics of power. The question for the cases is: if negotiations can actually lower transaction cost even in situations where uncertainty persists, what role does power play concerning the development of common solutions between actors of different size?

---

[184] See Newbery, 2000.

Essential facilities are a power source

Are transaction costs the reason for perceived power in negotiations? Does perceived power originate from essential facilities and 'big' vs. 'small' situations in network markets?

Transaction costs differ according to the access to the network. Different actors in network markets have different transaction costs. This is a function of the characteristics of the network markets: some actors have free access to the facilities (and the right to use them) and others also need them in order to be able to offer their services in the first place.[185] In this respect the business model of one actor depends on his access to the facility, but the business model of another actor does not depend on the sharing of the facility.[186] As the actors strive for greater product efficiency, they try to gain access to new technologies and to innovations.[187] This leads to a need for negotiations but with an asymmetrical distribution of power between the actors. The fact that there are cases where one company sues another to receive a great amount of money as compensation because the other has 'abusively aborted negotiations'[188] shows that cost savings are expected and are the major driver for negotiations.

Lobbying in network markets mirrors the principal agent theory. In the theoretical chapter of this study it was argued that the principal agent theory connects the different levels of access which actors have to information and also their ability to process it. It differentiates, furthermore, between those actors who negotiate as an individual and those who negotiate as a company. For the network markets this can be demonstrated by considering the lobbying efforts of the actors. The so-called 'logic of lobbying'[189] describes the situation of lobbying as follows: actors offer information and expect decisions that favour them in return.[190] Lobbyists simply do not have enough information to know where and how to best approach a lobbying situation. Thus, especially in such a dynamic political environment as the network sectors, those that offer information cannot be sure that it will be used in the way they originally intended. The connection between input into and output from the lobbying process is non-

---

[185] See Canoy/ Bijl/ Kemp, 2004.
[186] See Hirschfeld, 2003; Scheurle, 2003.
[187] See Heng, 2004b.
[188] See FAZ, 2005g, p. 18.
[189] See Michalowitz, 2004, pp. 17ff.
[190] See Miller, 2003, pp. 76 ff., for examples in the rail sector; EON, 2005, for an example in the electricity market.

transparent.[191] Therefore lobbying cannot be managed as the principal would hope in network sectors and does not deliver the certainty for which he strives.[192] Some argue that only stronger actors can afford the cost of lobbying.[193] The cost of lobbying, a part of the total transaction cost, varies according to the size of the actors, who range from individuals to companies.[194] Simply by virtue of this fact different actors have different transaction costs in the same negotiation situation.

It seems to be a typical feature of network markets that associations handle a large part of the lobbying as smaller and bigger actors compete for the same market.[195] This is especially true for newly liberalised markets.[196] According to a study from Putz & Partner[197] analysing the ten major associations in the telecommunications market in Germany, their efficiency is considered low. The associations bundle the interests of their members but mainly seem to lobby for themselves. Even though the awareness of these lobby-groups in the relevant institutions exists, they are not participating in the core decisions that affect their members. In the German network markets it seems as if there are only a few associations that know how to handle power to the advantage of their members.[198] Instead many actors, even though they belong to one or more associations, have chosen to take matters into their own hands or work with public affairs agencies.[199] This also points to the principal agent dilemma which the actors find themselves in: associations and companies depend on the individuals who execute their lobbying.[200] Even seemingly homogenous groups, i.e. the associations that strive to get more attention in the media, in politics and with the regulator on telecommunications issues, can develop different interests, even if that is consonant with splitting up the group. Where individuals act in this way the force of such a group is considerably weakened and it is likely that the 'small' companies thus diminish themselves.[201] In the German telecommu-

---

[191] See Neuhoff, 2003; Werner, 2002.
[192] See Cassel, 2001, p. 112.
[193] See Evans/ Fingleton, 2002, p. 15; Strauch, 1993, p. 7.
[194] See Althaus, 2004, p. 8.
[195] See Strauch, 1993, pp. 147 ff.
[196] See Schönborn/ Wiebusch, 2002, p. VIII.
[197] This paragraph builds on Putz & Partner, 2004.
[198] See Baethge/ Hübner, 2004, for an overview of the efficiency of the German lobby-associations in general.
[199] See Schönborn/ Wiebusch, 2002, pp. 93 ff., for a specific example of a campaign run by the internet-provider AOL.
[200] See Rochlitz, 2005, for the difference between personal and institutional influence with examples.
[201] See Werner, 2002; Heinrichs, 2004; Scheurle, 2003.

nications sector the association of the new competitors (VATM) for example is split into eight different interest-subgroups. These subgroups have different opinions on the same issues, for example, on how, if at all, to regulate the mobile telecommunications sector. This missing focus does definitely not make the association more credible nor its pronouncements more univocal.[202] As the telecommunications market has become more and more an international market, especially in regards to the regulation, the missing international activities of the associations endanger their efficiency. All in all the study of Putz & Partner concludes that the cost benefit relation of the lobbying activities is rather critical.

At first sight this seems to indicate that lobbying is inefficient for the actors in both of these network markets. Instead of creating harmony, lobbying could create more extreme positions.[203] It seems to be a fact that decisions can only be altered to one's advantage if those being influenced are at the very beginning of their decision-making process.[204] But combining the available resources and setting up competence networks[205] between the actors allows the actors to gain more certainty as more mutually advantageous options are created.[206] The handling of information is a bottleneck – an essential facility – in negotiations in network markets. Despite the problems, lobbying in network markets offers potential for making issues more transparent to those involved in negotiations.[207]

Dealing with essential facilities in networks is called interconnection. The unique characteristics of network industries appear to set them apart from most other industries which are deemed to be 'competitive':[208] the production is very capital intensive and there is scant opportunity to store output. The variable costs are very low, economies of scale are eminent, and sunk costs are high and relevant. Positive external network-effects (the more users, the better) exist.[209] In the telecommunications market the cost of operation is independent of the flow of services; in the rail transport market low margins, high investment and a slow growth function act as a barrier to market entrance.

---

[202] See Scheurle, 2003.
[203] See Epstein/ Nitzan, 2005, pp. 15 ff.
[204] See Scheurle, 2003.
[205] See Althaus, 2004, p. 7; Putz & Partner, 2004.
[206] See Freytag, 1998; Soltwedel et al., 1986, pp. 284 ff.; Becker, 1983 and 1985; Fockenbrock, 2005, p. 24, on the explicit example of DT.
[207] See Strauch, 1993, p. 8; Schönborn/ Wiebusch, 2002, p. 11. For a critical view see Grass, 2005, who states that lobbying as such is a danger for society.
[208] See Glachant, 2002, p. 297.
[209] See Stigler, 1971.

3.2 Power in network market negotiations

During the process of network markets liberalisation the problem of essential facilities came under the regulators' spotlight,[210] the latter having noticed that negotiations almost always stopped when it came to essential facilities.[211] In this sense the Essential Facilities Doctrine is relevant for network markets.[212] Bottlenecks could be the – or at least a – cause of asymmetrical power situations in network markets.[213] The existence of essential facilities, and the problems associated with them, are enshrined in, for example, the German telecoms and cartel law.[214] In respect to the network markets, the discussion of essential facilities and the regulatory need to intervene can be subsumed under the headings of price and quality. Because politics make negotiations in network markets even less transparent it seems almost impossible to design a sound regulatory regime which appreciates both aspects.[215] At least until now regulatory concepts have seldom accomplished these two tasks at the same time.[216]

The handling of the essential facility phenomenon 'translated' into the practical negotiation situations both in the telecommunications and in the traffic market is called interconnection.[217] Interconnection in technical terms describes the act of connecting the different parts of networks and related services so as to make it possible for the user of telecommunications services to call from a certain point to another point (any-to-any connection). In economic terms the difficulty of interconnection lies in defining what exactly is essential vis-à-vis the interconnection services and thus which services need to be regulated and to what extent.[218] Which products or parts of products are considered to be an essential facility?[219]

---

[210] See Knieps, 2004, pp. 18 ff.
[211] For both markets examples of the misuse of power in connection with essential facilities exist. For the telecommunications market see VATM, 2001. For the rail market see Mehr Bahnen, 2003 and 2005; McKinsey, 2002. Kirchner, 2003 and 2004 list all the theoretical market barriers for rail markets and structure them.
[212] See Heinen, 2001; Peltzman/ Winston, 2000; Vogelsang, 2002; Wissenschaftlicher Beirat beim Bundesminister für Verkehr, 2002. See PSPC, 2004, for a listing of the essential facilities in the German rail market.
[213] See Engel, 2002b.
[214] In the German Cartel law (GWB) this is regulated in § 19 (4) No. 4.
[215] See Bijl/ Peitz, 2002, p. 243.
[216] See Scheurle, 2003; Heinrichs, 2003.
[217] See Shapiro/ Varian, 1999, p. 2.
[218] See Die Bahn, 2004, p. 17, for the point of view of an incumbent on essential facilities in the rail sector.
[219] See Kirchner, 2003.

The most prominent interconnection cases have been (just) about price issues.[220] The competitors claim access to everything the incumbent offers on a cost basis.[221] In the rail transport market this is illustrated by discussions surrounding the setting of the price for track usage;[222] in the telecommunications market it is the discussion about the level of the interconnection prices.[223] The case of the introduction of end user flat rate internet tariffs in the telecommunications sector illustrates the fact that an interconnection negotiation is not only about price but also about the quality of the service. The incumbent DT offered such an end-user flat rate tariff in 2001 without providing corresponding services for the competitors as a pre-product.[224] When the competitors asked for the corresponding pre-product, Deutsche Telekom offered none, but pointed to the minute based tariffs which they already provided to their competitors on the basis of their interconnection agreements. Incumbents the world over responded in a comparable manner. The regulator was called in, but it was only in 2004 that the TKG 2004 explicitly states that different tariffs are to be seen as different essential services (not only as different prices for the same product) which are to be provided by the incumbent on an interconnection basis.[225]

In the German rail transport sector the discussions regarding the use of the maintenance facilities of the incumbent by the competitors did not even get as far as the question of price.[226] It would seem rational, though, that the incumbent acquires new maintenance contracts as the new rolling stock has fewer maintenance intervals and thus he himself no longer operates his maintenance facilities at full capacity.[227] But the incumbent was at first unwilling to even start negotiations with his new competitors with a view to letting them use these facilities. Another such case is that the rail incumbent in Germany does not sell off old rolling stock but prefers to save it as a stock of spare-parts.[228] The competitors even claim that the incumbent 'hides' this material in order to make absolutely sure that no one else knows what is available.[229] Since the rolling stock is cost intensive and the suppliers have long delivery times, a shortage for the whole competitive

---

[220] See Freytag/ Winkler, 2003; Ewers/ Ilgmann, 2001a, p. 6.
[221] See Rochlitz, 2003; Neuhoff, 2003.
[222] See Knieps/ Brunekreeft, 2000.
[223] See Monopolkommission, 2004a; RegTP, 2003b.
[224] See Vanberg, 2002, pp. 8 ff., for the details of this case.
[225] In the German TKG 2004 this is regulated in § 27 (3).
[226] See Die Bahn, 2004, pp. 34 ff.
[227] See Hirschfeld, 2003.
[228] See Die Bahn, 2002b, pp. 31 ff.
[229] See Rochlitz, 2003.

market is provoked by such behaviour on the part of the incumbent. Further examples of (conflict-) negotiation issues on essential facilities exist in all network markets:[230] the many different details regarding the cooperation of the different traffic and infrastructure providers, the availability of the rolling stock, the access to timetables and the use of the infrastructure, i.e. the network itself.[231] Are a harbour and its facilities a part of the service between different cities, or can anyone build their own facilities in a relatively small harbour?[232] Even the access to (rail-)energy was one of the conflict issues: the German court ruled that it was part of the essential facilities as there was no other supplier of that special kind of energy[233] and DB was obliged to give its competitors access to its energy networks.[234]

For all these cases the question was whether a certain product is to be considered an essential facility which has to be provided by the incumbent. The incumbent calls such services 'extra-services' to indicate that, from his point of view, they are not essential and that he thus sees no need to provide his competitors with such services.[235] But even access to essential facilities alone does not guarantee non-discrimination. Non-discrimination is achieved when the incumbent treats himself no better than the other actors. Non-discrimination is not realised if the incumbent discriminates against all the competitors in the same way as the following case of the 'missing splitters' for DSL connections in Germany demonstrates: a truck fully loaded with splitters which are used to set up the broadband connection at the customers home and which were supposed to be delivered to the competitors of Deutsche Telekom had an accident. The competitors of DT were not able to set up connections for their own customers for 10 days, at the same time Deutsche Telekom had no shortage.[236] The difference between access and non-discrimination can be illustrated with reference to the following cases from the network markets. A large choice of leased lines in the telecommunications market are available for all kinds of connections to all carriers;[237] there even seems to be an overcapacity of such lines and the services available on them. But when one looks at the difference between the supply and the quality of the supply, one notices that

---

[230] See Schweinsberg, 2002, for a listing of the different cases. 75% of those revolve around issues of access.
[231] See Heinrichs, 2003; Rochlitz, 2003; Hirschfeld, 2003.
[232] See Hirschfeld, 2003; Heinrichs, 2004.
[233] See Krummheuer, 2005a, p. 17.
[234] See FAZ, 2003b, p. 12.
[235] See Die Bahn, 2004, pp. 34 ff., for a discussion of these 'extra-services'.
[236] See Petersdorff, 2005, p. 33.
[237] See RegTP, 2002, for this case.

certain routes and services are only available with a very long delivery time, which even then is not guaranteed. Incumbents tend to sell those lines which are competitive with short delivery times and for very low prices, and those that are not with long delivery times and for a high price.[238] This could be a sign of the use of asymmetrical power. An example from the rail transport market is the competitive operator's request for a track that crosses some of the most frequented lines of the incumbent. The incumbent assigns track-times, but the quality (in terms of timing and stops) is not the quality that the operator asked for; not only does the incumbent get the assigned lines, but also a better quality.[239] Discrimination, it is feared by regulators, could result in actors leaving the market and thus endangering the whole process of liberalisation.[240] The term non-discrimination is hence closely related to power: those who have access to an essential facility, or even better total ownership of it, have many possibilities to discriminate, using the scarcity of the essential facility as a threat or as an instrument of power. Access to broadband is a politically highly sensitive issue.[241] Those regions that for technical reasons do not yet have access to DSL broadband by the incumbent feel left behind. Therefore the approach of the incumbent Deutsche Telekom to ask those regions for a subvention to get DSL coverage, was the demonstration of power: if you want it, you have to pay extra. The competitors complained quickly as they saw their niche markets being taken over by the incumbent.[242]

Separating network and operations could ensure that no one actor has better access to the essential facilities.[243] The EU liberalisation reforms of the early 1990s were aimed at unbundling infrastructure and operations and opening access for competitors[244] in order to increase efficiency in these markets.[245] Both the telecommunications market and the rail transport market are network markets which can be disaggregated into three lev-

---

[238] See Scheurle, 2003.
[239] See Die Bahn, 2003; Rochlitz, 2003.
[240] See Scheurle, 2003; Neuhoff, 2003.
[241] This paragraph builds on Bundesministerium für Wirtschaft und Arbeit, 2005.
[242] See Berke, 2005, p. 10.
[243] See Winkler/ Rochlitz, 2005; BDI, 2005, pp. 8 ff.
[244] See Die Bahn, 2002b, p. 7. Even the incumbent DB admits that the crucial characteristic of a competitive market is the open access to it. Obviously, different actors have different ideas as to what 'open' means.
[245] See OECD, 1998, pp. 7 ff., for the theoretical discussion on alternative models of ownership and control in the rail transport markets and an extensive overview of 15 countries. Here only the German and the UK market will be considered in greater detail. For a critical view that the separation has a positive effect on regulatory reform, see Friebel/ Ivaldi/ Vibes, 2004, p. 4.

els:[246] the traffic level, the management level and the infrastructural level.[247] Both markets combine within them the control and management of service and infrastructure. Other transportation markets have control over either one, but not over both at the same time.[248] Those that own the essential facility, even where they are not the sole owner, can profit more easily from the economies of scope, i.e. the advantages that the combination of the different parts of the network offers. The incumbents even emphasize this fact, but in terms that defends the idea of one network[249] in the hands of a single owner.[250]

The question is how the transaction cost of negotiating essential facilities can be limited by organising the bottleneck so that it can be efficiently used by the different actors.[251] The ability of the competitors to negotiate access to the scarce resource network could be improved if an institutional arrangement, such as unbundling, were to make sure that the actors negotiate themselves. Only if there were disagreement would the regulator step in.[252] Both for the telecommunications market and for the rail transport market such an arrangement seems to be have been developed furthest in the UK, where the British rail transport system was completely privatized.[253] To limit the supplier's dependency on the incumbent,[254] the UK approach was to split up the incumbent British Rail into private enterprises: Railtrack (RT) runs the network and the stations, the Rolling Stock Companies (ROSCOS) the rolling stock and the private train-operating companies (TOCs) run the services. These companies work together on a negotiated contractual basis. The Rail Regulator (RR) and the Strategic Rail Authority (SRA) regulate this regime on non-discriminatory terms. The pool of rolling stock in the UK is provided by leasing companies to all providers on a non-discriminatory basis.[255] The first attempts at reforming

---

[246] See PSPC, 2004, pp. 46 ff.
[247] See Knieps, 2004, p. 2.
[248] See for example the air transport sector (traffic or airports).
[249] There is actually no such thing as 'one network', but only many small parts that are connected to become a network. The main idea of a network is to extend its reach constantly.
[250] See Hirschfeld, 2003; Die Bahn, 2002b.
[251] See Hirschhausen/ Siegmann, 2004.
[252] This paragraph builds on Héritier, 2001b, pp. 6 ff.
[253] See Lavender, 2005. In the telecommunications market the incumbent BT is (in 2005) about to set up a separate 'Access Service Division' with its own headquarters and brand but which will remain part of BT.
[254] See Böhmer/ Schnitzler, 2003.
[255] See Die Bahn, 2004, pp. 20 ff.; Die Bahn, 2005b, p. 39. The incumbent DB is obviously against such an approach and voices its opposition loudly.

the system failed in 2001, however, when after several accidents RT was again placed under state control, the RR was shorn of major responsibilities and the ministry took over most of its duties.[256] Current developments in the discussion regarding the separation of network and services in the German rail transport sector are marked not so much by economic, but by political considerations as to the feasibility of a forthcoming IPO of the incumbent Deutsche Bahn, with or without its own network.[257]

The discussion of the cases shows that essential facilities are a source of power, which creates asymmetries and the need for negotiations. The (conflict-)negotiations centre around the essential facilities, where there is no other source of service provision besides the incumbent. Access to essential facilities is not only about the price, but also about quality – both of which have to be dealt with simultaneously.[258] But obviously at the same time many regular, non-conflict negotiations need to take place as well, in order to satisfy the day to day demands of the actors.[259] Altogether the distribution of power between the actors seems to be a decisive factor in the cost equation of gaining access to the essential facility.

Sometimes 'big' actors are 'small' and 'small' actors are 'big'. Network markets exhibit network effects that result in perceived advantages for 'big' actors: the bigger a network, the more scale effects can be realised.[260] Smaller networks are then not as attractive to customers. If the actors do not make the positive external network effect available to their customers, these might not appreciate the service and stop using it altogether.[261] This reveals a dependency on the part of the providers insofar as they need to interconnect their networks. Whereas the user sees just one big network which connects all, the providers have to connect the different small networks so as to provide a single network for the customers,[262] causing transaction costs in so doing.

---

[256] See NZZ, 2004.
[257] See for the current discussion Allianz pro Schiene, 2004; FDP, 2004; Ewers/ Ilgmann, 2001b; BDI, 2005, pp. 8 ff; FAZ, 2004c.
[258] See Ewers/ Ilgmann, 2001a, p. 6. They look at the pricing system of Deutsche Bahn, which discriminates between the competitors: not so much on pricing issues but on quality aspects. These are the issues that are hard to prove and can be extremely disturbing as they cause cost in terms of time and legal expenses.
[259] See Werner, 2002.
[260] See Köster, 1999.
[261] See Siedenbiel, 2003, p. 78. Connex took one of its main lines in Germany out of service, because the level of usage dropped. One of the reasons for the end of the service was the lack of connection with the incumbent's trains.
[262] See Schintler et al., 2004.

This is the reason why both network markets were considered to exhibit characteristics of a natural monopoly and why still today both markets are characterised by (but not limited to) incumbent vs. new alternative operator situations.[263] In most parts of the relevant markets the incumbents have high, even monopolistic, market shares. This also means that the 'big' actors might have an advantage, since they have most of the customers on their network already.[264] Additionally they have very long running customer relationships, i.e. their customers are not willing to change from their long term supplier to another newer one.[265] The acquisition of customers by 'small' companies is thus much more costly than it is for an incumbent.[266] Incumbents are large companies (they are, for example, among the largest employers in Germany[267]), and were formerly state owned bureaucracies. On the one hand this gives the 'big' actors more resources for establishing and maintaining power in negotiation situations. On the other hand being 'big' is a costly business. Firstly, because they have more customers, 'big' actors have more to lose as well, responsible as they are for dictating trends within the market. The internal structures of the incumbents in the network markets are inefficient and open to political influence.[268] The many different levels of hierarchy within a 'big' company make it necessary to involve and harmonise many interfaces, which in turn makes the decision-making process as such extremely complicated.[269] The larger the company, the more difficult and expensive the negotiations are,[270] as there is a pressing need for coordination which makes the process of fine tuning time consuming.[271] For example, the incumbent Deutsche Telekom started out with a rather small interconnection department at its headquarters in Bonn when the liberalisation started. In 1999 the organisation was regionalized so that the customer contacts were closer to where the customer was located. But the bottleneck within the organisation is still the headquarters in Bonn, since every action in the region has to be ap-

---

[263] See Scheurle, 2003; Neuhoff, 2003.
[264] See Bundesministerium für Wirtschaft und Arbeit, 2003; Schäfer, 2003, gives a detailed telecommunications market overview from the perspective of the consumers.
[265] See Werner, 2002.
[266] See Neuhoff, 2003; Rochlitz, 2003.
[267] See FAZ, 2005h, for a listing of the TOP 100 companies in Germany.
[268] See A.T. Kearney, 2004.
[269] See Hirschfeld, 2003; FAZ, 2005e, p. 19.
[270] See Wünschmann, 2005; Scheurle, 2003.
[271] Die Bahn, 2004, p. 1, points to the complexity of the day to day work in respect to the changes necessary when working with many new, small companies – a process which is caused and controlled by regulation.

proved by the headquarters first before it is put into effect. This slows the decision-making process down. Additionally, the incumbent is obliged to get every product approved – or at least he has to make sure that it will be approved – thus prolonging launch times.[272] 'Big' actors also need to react on different levels at the same time. Whereas they might be 'big' on the national level, on the international level they are confronted with the same problems that other 'small' operators are battling against on the national level. For example DB is complaining about the hold-up in negotiations on the international level with other international operators.[273] Altogether this minimises the effect of being 'big'.[274] 'Small' actors are fast and more flexible,[275] a fact which is also true for network markets. In terms of innovation it is mostly 'small' companies that are responsible for launching new product variations.[276]

This 'indecisiveness', regardless of whether an actor is 'big' or 'small', can be illustrated with the following case: the first interconnection cost regime for the liberalised telecommunications-market in Germany was set by the Federal Ministry of Post and Telecommunications in 1997. The regulator set a minute price for the services of the incumbent DT on the basis of a benchmark of other European telecommunication operators, because the negotiations between the actors had failed to come to a conclusion.[277] What was not explicitly mentioned was that this price was to be used in 'both directions', in other words the so-called interconnection price was not only valid for the services provided by DT but also for those services provided by DT's competitors. This came to be called 'the concept of reciprocity'.[278] The concept was thought to be correct and fair, because the (economic) reasoning behind the regulated pricing was the underlying cost structure (whose details were initially unknown, hence the benchmark); moreover the price was to reflect an efficient provider of telecommunication services.[279] But with the introduction of the EBC regime in 2001 the concept of reciprocity began to be questioned for the first time. It was DT that pro-

---

[272] See Höckels/ Rottenbiller, 2003. The incumbent feels discriminated against in terms of his ability to introduce innovative products.
[273] See Balsen, 2005, p. 10.
[274] See Hirschfeld, 2003; Wünschmann, 2005.
[275] See Frühbrodt, 2002a p. 15.
[276] See Monopolkommission, 2004b. See Pergande, 2004, p. 18, for an example of such a product variation.
[277] See Schütz, 2005a, p. 166.
[278] See IRG, 2005, for an international overview of the concept of reciprocity.
[279] See Ruhle/ Lichtenberger/ Kittl, 2005, p. 10, who describe that the concept of reciprocity is also seen as a basis of the economic liberalisation principles in Austria.

posed to pay different prices for different networks, i.e. lower prices for smaller networks than those of DT, arguing that those were more efficient as they were new, in contrast to the 'old' network of DT. At that time the regulator stopped this line of argument against the concept of reciprocity by ordering one price for the interconnection services, but only if the operator agreed to expand his own network as well once a certain level of utilization was reached. Thus RegTP did not discredit the concept of reciprocity as such but connected it with a restriction on new operators by requiring them to invest in infrastructure in order to gain access to a lower interconnection price. In 2003, though, a large number of city carriers, i.e. 'small' actors, asked the regulator to allow them to charge higher prices than those set for the incumbent for the usage of their networks. RegTP conducted a public hearing.[280] The feedback was very mixed.[281] Despite this, the regulator ruled in favour of the city carriers, ending the pure reciprocity regime and establishing asymmetrical pricing. Basically this case is about cost effects in the network market and the possibilities of negotiations when power is distributed asymmetrically between the actors.

DT is a 'big' incumbent and its organisational development has a long history. The organisation is large, carries old burdens (pensions etc.) and is constantly being restructured. The cost of extracting the necessary cost data in preparation for a negotiation is high. On the other hand, the organisation has plenty of employees who can do the necessary organisational work, splitting the work into manageable chunks for those specialists that can handle it most efficiently. The (network-) infrastructure of DT is the most extensive in Germany as it has evolved over a long time period. Not all the equipment is the most modern, but on the other hand much that which is still in use has already been written off and is hence very cheap to run. If changes need to be applied to the network as a whole, the cost of these changes might sometimes be high, because one is dealing with a large network, whose many different functions react differently to changes. But a large network also brings advantages with it: if parts of the network go down, the traffic can still be rerouted without loss of service-quality.

'Small' competitors, such as the city carriers, are mostly outsourced from formerly larger institutions (savings and loan banks, public services) that employ fewer people, but seem to have a higher output as they only add customers to their network if they expect them to be (highly) profitable. Most of their network infrastructure is 'brand-new'. The cost of the infrastructure is greater because of modern technology and the higher ser-

---

[280] See RegTP, 2003a.
[281] See Williamson, 1993, pp. 45 ff. Neumann, 2004, p. 3, discusses reciprocity as a (necessary) part of a consistent pricing regime.

vice cost, but it is at the same time more efficient. Output per unit is generally higher, but then again it is more prone to breakdowns. Since the overall network is smaller, there are more bottlenecks that can hold up the service as a whole, and these cannot be made up for through sheer size. Much of the technology used is not 'experienced' (for example the billing system) and neither are the operators. Scale and scope effects are scarce. The only time the scale effects are larger than those of the incumbent is when a whole new district is being added to the network, generating in itself a small local monopoly.[282] Thus the smaller competitors have less (cost-) data, but also fewer employees to handle it. They take less time to decide and hence operate faster but are not as powerful when it comes to assembling the data.[283]

When actors (of different size) negotiate, they reveal that they have a different cost structure and a different sensitivity to perception of time: 'small' carriers fear a time lag caused by the 'big' carrier,[284] as incumbents prefer to leave everything as it is. That power is distributed asymmetrically between the actors hence seems to be the normal rather than the exceptional state of affairs. It is, therefore, 'normal' in negotiations for one actor to have higher transaction costs than the other, though how this affects the outcome is unknown.

Résumé: Complexity of essential facilities causes transaction costs. Essential facilities are a source of **perceived** power in negotiations which cause transaction cost as there is a need to negotiate the access to them and as actors are trying to minimise their cost (of conflict). In confirmation of **Hypothesis 4**, it can be stated that transaction costs cause asymmetries and are a source of uncertainty. It is the complexity of essential facilities that creates an asymmetrical distribution of power in negotiation situations in network markets. Not only might there be a 'right time' to negotiate certain issues,[285] but there also seems to be something like a minimum size that actors have to achieve – possibly with the help of associations, lobbying or the regulator – if they are to be respected at all in negotiations. They need to grow before they can become a David first. Only then can they slay Goliath.

---

[282] See Commission of the European Communities, 2005a.
[283] See FAZ, 2005d, p. 13, for a description of the same effect in the rail sector.
[284] See Rochlitz, 2003; Werner, 2002.
[285] See Hirschfeld, 2003; Rochlitz, 2003; Werner, 2002.

Uncertainty causes dynamics of power

The power derived from the access to essential facilities gives some actors – at least initially – a feeling of security. This illusion of control does not last long, though, because the dynamics of power create uncertainty. It then becomes less obvious who is powerful and who is not.

Power is only an illusion as actors in network markets ultimately depend on each other. Actors do not know if they are powerful or not in a certain negotiation situation.[286] This seems to have something to do with the way actors make decisions as an individual or as a company and in due course generate transaction costs. There is, more than theory would initially seem to suggest, a significant difference between situations where an individual or a company is negotiating.[287] Even in negotiation situations where it seems to be obvious that a 'big' actor and a 'small' actor are taking part it is still the individual actors that run the negotiation. The following case may demonstrate the importance of individuals in negotiations: to implement the so-called 'pre-selection' (dial-around) regime in the German telephone network, the incumbent must apply changes to his own network every time an end-user places a pre-selection order. But it is the new operators that have to inform the incumbent of those changes. Since it is the incumbent receiving these order forms, he has been able to define their format.[288] The quality of the implementation of these orders depends greatly on the personnel on both sides. When the incumbent started to derogate the quality of the implementation process, the new operators soon followed with counter reactions. It seems as if the new operators were trying to hurt the incumbent in order to show that they, too, were able to make the implementation process more unreliable. These measures were not aimed against those executing the strategy (the individuals), but were trying to weaken the position of those who had formed the strategy (the company).[289] In the German rail transport market the incumbent DB publicly blames individuals within the competitor Connex for the souring of the relationship.[290]

These examples show that negotiations depend on the individual actor.[291] The different speeds at which actors work and the different issues that they care about point to the different levels on which they operate.

---

[286] See Hirschfeld, 2003; Rochlitz, 2003.
[287] See Winkler, 2003.
[288] See Tele2, 2005, pp. 27 ff., for this example.
[289] See Scheurle, 2003.
[290] See Rochlitz, 2005.
[291] See Heinrichs, 2004; Hirschfeld, 2003.

This can be illustrated with the following case: Deutsche Telekom negotiates its extensive network interconnection agreement with more than 100 different alternative operators. The new operators have diverse business models (local, long distance, mobile business, fixed network business, small, medium, large, national and international customers and any combination of these). The incumbent operator tries to minimise his expenditure in terms of negotiating time by limiting them all to a standard interconnection agreement (RIO). It could, at the same time, be worthwhile for the incumbent to strike different deals with the different operators and thus exercise price discrimination, but the regulator only allows for one non-discriminatory price for the same service.[292] It is difficult to state who has more power in such a situation. It could be the incumbent, because he is the one who decides which services the operators (and there are more than a hundred of them) will actually gain access to. It could be each and every one of these new operators, because they can always blame the incumbent when something went goes wrong, and make accusations against him to the regulator (who is likely to believe them). It could be the regulator, because he has the power to set the prices with a ruling, even though he is not fully informed of all the details needed for a profound economic, legal and technical decision.

Reality shows that it is not just a single decision which is made by the actor on how to handle a special situation as a company, but that many decisions are made by the different individuals in their individual negotiation situations. The following example involving the incumbent Deutsche Telekom might demonstrate this: from 1996 until 2002, when Ron Sommer was CEO of DT, his clear message to competitors and politicians was that monopolist power was being put to aggressive use. When Ron Sommer left office a new, less aggressive style of management was announced and the managers who represented this aggressive style to the outside world were subsequently removed.[293] The question is whether actors who need to negotiate with the incumbent find that not only the 'company DT' but also their individual counterparts have changed their kind of behaviour. This example demonstrates the fact that one cannot expect every individual in the company 'to be always on message' regarding company 'propa-

---

[292] See Stoffels, 2004, p. 6, who describes that in practice the incumbent can even use the non-discrimination clause against the new operators, when he is not willing to give access to a new product or price.
[293] See Riedel, 2002.

ganda'.[294] That is why each and every interconnection agreement is different from the others, based as it is on the individual negotiation. The inability of large actors to master their own actions completely could make 'small' actors more powerful than has been suggested: it is not a 'big' company against a 'small' company at the negotiating table, but individuals. The cases show that uncertainty exists in the negotiation situation about the actual power distribution, even if it seems obvious, at first, that a 'big' and a 'small' actor are facing each other.

In an essential facilities market actors depend on each other, but it is not obvious who depends on whom or to what extent. If only one actor depends on the other, this is called a 'one way essential facility'.[295] In this case the power position is better for the independent actor. If both actors depend on each other, those are 'two way essential facilities'. The difficulty lies in defining whether an actor is independent or not, as can be illustrated with the following example: Connex, the largest German competitive operator, complains about the poor image of the incumbent Deutsche Bahn because the effect of this image is felt by all rail transport operators.[296] This shows that one cannot be sure if one is not in one way or another dependent on the other actor.[297] This could put the actors on the same level, in terms of power, as both depend on something provided (only) by the other. Essential facilities would then no longer automatically be a source of power for just one actor. The interdependency of the different parts of the network, without which the original essential facility could no longer grow or even maintain its attractiveness, places the different actors on the same level, especially in single issue negotiation situations.[298] It can be concluded that in negotiations on interconnection issues the climate between the actors is marked by uncertainty, i.e. it is unclear who has the power of the moment and who depends on whom in order to achieve what.[299] The resulting cost basis is thus also marked by uncertainty.

Uncertainty might be a cost advantage for 'small' actors. The problem with power is that the actors cannot determine exactly what influences what in a negotiation situation, nor what a particular transaction cost is

---

[294] See Wendel, 2004, on the internal problem within DT of implementing this reform. See Tartler, 2005, for an example of DB trying to keep up the propaganda; Neuhoff, 2003; Heinrichs, 2003.
[295] See Armstrong, 1998; Scheurle, 2003.
[296] See Grabitz, 2003.
[297] See Wünschmann, 2005.
[298] See Scheurle, 2003; Hirschfeld, 2003.
[299] See Rochlitz, 2003; Hirschfeld, 2003.

caused by. Sometimes the effect of being 'small' is equivalent to that of being 'big'. 'Small' mobile operators, for example, may have more power than large operators as they do not influence the perceived total average mobile price as much as the 'big' carriers and can thus charge higher prices without being punished by the customers.[300] Additionally, by making use of their lower cost base, smaller operators can become 'big' in terms of power in negotiations.[301] They can organise their power better by bundling the interests of several 'small' actors so as to command the attention of those that regulate the market.[302] Differences in power can eventually be compensated for by increased activity.[303] In 'fast' markets, especially, rapid mechanisms are necessary to cope with conflicts and special market conditions. Such mechanisms as the internet and electronic media, deliver – in spite of their low cost – the essential support which enables 'small' actors to become 'big'.[304]

'Small' operators, for example, can disturb the incumbents by ruining the latters' image. In the rail transport sector in Germany there is, for example, the user-group Verkehrsclub Deutschland (VCD), which on a regular basis tests the Deutsche Bahn and publishes the results of its findings.[305] It is on the basis of this test that the VCD formulates its demands. With even a modest investment the effect of an attack on a large company can then be fairly 'big'. Such a power set-up leads to a different starting position in negotiations for those that were originally 'small'. This shows that the special characteristics of network markets make the networking of the actors an integral part of the power play in negotiations.[306]

Does the regulator determine who is powerful? The change from a monopolistic network market to a liberalised market causes costs for the market players since they have to adapt to the changing environment.[307] On the one hand, actors have an incentive to find the right amount and the right price for the essential facilities in negotiations,[308] without the interference of a regulator. On the other hand, without the support of an external party it is difficult for the actors to even establish a basis for negotiation in net-

---

[300] See Arnback, 2005.
[301] See Werner, 2002.
[302] See Rochlitz, 2003; Hirschfeld, 2003; Werner, 2002.
[303] See Michalowitz, 2004, pp. 229 ff.
[304] See Märker/ Trénel, 2003, p. 7.
[305] See VCD, 2004b.
[306] See Köster, 1999, p. 168.
[307] See Hirschfeld, 2003; Werner, 2002; Wünschmann, 2005.
[308] See Schneider, 1995.

work markets.³⁰⁹ Incumbents in particular are willing to defend their position, even if it is clear that such a position cannot be maintained for a long time.³¹⁰ To limit the cost of conflict and of negotiations in interconnection cases, it could thus be advantageous if a regulator supports the process.³¹¹ But as regulators regulate asymmetrically in order to grant non-discriminatory access to essential facilities, the actors have an incentive to gain additional power from the regulator.³¹²

For 'small' actors it could be easier to get attention from the regulator as they represent a minority and as such are more likely to enjoy the protection of those institutions responsible for overseeing the markets.³¹³ Connex, for example, paints a picture of a 'David versus Goliath' situation in its relationship with Deutsche Bahn in order to gain more support from the regulator.³¹⁴ On the other hand, however, regulation tends to make the uncertainty as to who is 'big' and who is 'small' in respect to transaction costs even worse, as the following example shows: in the world of the Internet-protocol (IP) all operators 'peer' (peering is the interconnection of the IP-operators), regardless of what size they are and even though the operators are aware that negotiations are not without cost.³¹⁵ There even seems to be a tendency towards excessive peering in the IP-markets. The difference between this and interconnection is that the actors can choose between different peering regimes as no regulation exists. Actors of similar size tend to peer (for free), others choose to interconnect via or together with other transit partners.³¹⁶ Being 'big' (because of regulation) is not necessarily equal to being powerful in a negotiation.³¹⁷

Lobbied as he is from all sides, the problem for the regulator is to find the 'right' instrument in the light of the existing power distribution and the need to create a non-discriminatory environment.³¹⁸ The problems that the regulators are confronted with are steered by multiple laws, ordinances and rulings. Regulating just one of the many issues might endanger the whole balance of the negotiation setting. That is why regulators seem to be de-

---

³⁰⁹ See Engel, 2002b, p. 18.
³¹⁰ See Heinrichs, 2004.
³¹¹ See Rochlitz, 2003; Neuhoff, 2003; Heinrichs, 2004; Scheurle, 2004.
³¹² See Neuhoff, 2003.
³¹³ See Rochlitz, 2003; Hirschfeld, 2003; Neuhoff, 2003.
³¹⁴ See FAZ, 2003a, p. 11; Rochlitz, 2005.
³¹⁵ This case builds on Jahn/ Prüfer, 2004.
³¹⁶ See Donegan, 2005, p. 29, for more technical details.
³¹⁷ See Hirschfeld, 2003; Heinrichs, 2004; Scheurle, 2003.
³¹⁸ See Engel, 2002b. See Bundesministerium für Verkehr, 2001; Wissenschaftlicher Beirat beim Bundesminister für Verkehr, 2002; Monopolkommission, 2004a; RegTP, 2002 and 2003b, for a description of the possible instruments.

termined to stick to a certain line of rulings instead of looking at each individual case.[319] Certain issues which might be referred to the regulator are just too complex to solve.[320] This could be a hint to the fact that the regulator should not try to involve himself in such complex issues but should leave it to the actors to try to find a solution.[321] Besides intervening directly, the regulator also has the possibility of facilitating negotiations between the actors, as the following example shows: in the past the incentives intended to encourage the setting up of a so-called 'billing contract' between DT and its competitors failed.[322] RegTP had passed several rulings to the effect that DT was obliged to share its billing services to a certain extent with the new carriers. No other European incumbent had to offer such a service arrangement,[323] a fact that Deutsche Telekom always complained about. DT claimed that such an obligation would not act as an incentive for the new carriers to try to set up their own billing services. In return VATM complained that DT was exercising their monopoly of the customer contacts by trying to close down a whole market segment (the so-called 'call-by-call' market). Finally, the legislator threatened to rule on how to handle the billing issues. Since billing is such a complex issue, it became obvious to all operators that such a ruling would have damaged all market participants, as it would not have been able to go into sufficiently exhaustive detail. In the light of such a threat,[324] Deutsche Telekom and VATM set up a (self-regulatory) contract which specifies in detail which services will be delivered and at what price.[325] The revision of the telecoms law in 2004 has also taken up this clause and made it obligatory. The billing contract, obviously, has provided a constant and efficient negotiation basis for the actors; it has been prolonged and extended in 2005.[326]

The examples above show that the direct involvement of the regulator tends to enlarge the uncertainty of the actors in terms of time and risk, rather than delivering planning security. Content and time, the major cost drivers for the actors in negotiations, are not what the regulators are primarily focusing on. Supporting self-regulatory processes, though, can help the actors to establish more certainty.

---

[319] See Scheurle, 2003; Werner, 2002.
[320] See Kohlstedt, 2003b.
[321] See Heinrichs, 2004; Scheurle, 2003.
[322] This case builds on Freytag/ Winkler, 2004. See Ritter, 2004, pp. 195 ff., for a detailed description on the genesis of the billing contract.
[323] See Stüwe, 2004, p. 14.
[324] See Mechnig, 2004.
[325] See Deutsche Telekom, 2004.
[326] See VATM, 2005.

Measuring the cost of power in a dynamic environment is not a priority. Actors in network markets do not consider every issue to be worth negotiating.[327] One reason for this is that negotiations have different costs for different actors according to the different sizes of those actors. Since the cost basis and the result of the negotiation have different effects on every actor, each of them has to make an individual decision for himself. But even though cost effects resulting from and appearing in negotiations seem to be obvious to the actors, naming and quantifying them is far more difficult.[328] The following case demonstrates that the cost-benefit relation of a negotiation is different for every actor. The stake for a large operator in a negotiation on lower interconnection costs is higher than for the 'small' competing operator as the incumbent would most likely have to offer the same interconnection prices to all operators. Netcologne, a city carrier in the telecommunications market, called the regulator in 1999 to complain about the interconnection prices for traffic to the mobile carrier D2. Netcologne was not able to prove that sustainable negotiations had taken place. The complaint, therefore, was rejected.[329] It seems that the larger actor, here D2, had more to lose and had invested more in 'winning' the negotiation.[330] Netcologne was, so it seems, more interested in a fast timing, getting the negotiation over and done with as soon as possible. It was not until 2004 that another operator, 01051, again tackled the mobile operators on the question of a lower interconnection pricing.[331] This time the preparation of the 'small' operator 01051 was much more thorough and they achieved a qualified victory: the interconnection was ordered, but no ruling was given regarding the price.[332] In reaction to this the large mobile carriers agreed to lower their prices before the case was decided and before the regulator ordered them to do so.[333] Measuring power in this situation is almost impossible,[334] even though it seems sure that asymmetry has supported the use of power in this case. From an ex post point of view none of the actors profited from the use of power as the uncertainty about the situation still persists. Even today it is not clear how such a situation would be

---

[327] See Rochlitz, 2003; Hirschfeld, 2003.
[328] See Hirschfeld, 2003; Neuhoff, 2003.
[329] See Schütz, 2005a, p. 170.
[330] See Scheurle, 2003.
[331] See Piepenbrock, 2004.
[332] See RegTP, 2004a, p. 769.
[333] See Koenig/ Vogelsang/ Winkler, 2005, pp. 47 ff.; Bauer, 2004, p. 19.
[334] See Koenig/ Vogelsang/ Winkler, 2005. pp. 32 ff.

handled in the future and hence the actors have to operate with a certain degree of risk at all times.[335]

The problem is that even if absolute cost numbers of the production of power were available, this would not give actors an indication as to how best to make use of it in their individual negotiation situation. As the case shows, possible indicators for the cost-calculation are the (expected) results of the negotiation and whether or not the cost-benefit relation can be kept positive. But the case also demonstrates that actors are confronted with the problem that they have no experience of an identical individual negotiation situation upon which they can base their decision; neither from their own experience nor from that of some other actor. No one negotiation situation on interconnection issues is alike.[336] Neither is it intuitively apparent which alternatives need to be included in such an analysis. Thus, measuring their investment in power in a certain negotiation situation is limited to comparing different negotiation situations and comparing their results. The understanding of the use of power in such situations is then an issue of experience or know-how transfer. The transparency of comparable situations faced by other operators can be used as a benchmark for quantifying one's own cost-benefit relation. The alternative operators do this by handling the different cases via an association.[337] As the incumbent is part of almost all negotiations it is easier for him to gain the know-how from experience. The establishment of an 'indicator' is therefore costly, but is necessary in order to measure the cost of the use of power. This shows that the actors, depending on their size, profit differently from power in negotiations and the challenge for the actors is thus to decide in advance how much to 'invest' in creating power in their negotiation situation. The difficulties with the determination of the efficiency of the input of power could explain why taking into account the possible cost effects of their actions is not a priority for actors when preparing or executing a negotiation.[338]

Résumé: Reduction of uncertainty reduces cost of negotiations. Differences in the size of the actors cause varying transaction cost of negotiations. The problem is that the actors are uncertain about their position in the negotiation in relative or absolute terms. In confirmation of **Hypothesis 5**, it can be stated that actors only enjoy illusory control in negotiations. This uncertainty leads to a higher cost base for negotiations as a whole, as

---

[335] See Koenig/ Vogelsang/ Winkler, 2005. They demonstrate that even when the so-called market analysis for this market is concluded by all EU member states, the national regulators will decide how they interfere in the markets.

[336] See Hirschfeld,, 2003; Rochlitz, 2003.

[337] See Rochlitz, 2003.

[338] See Hirschfeld, 2003; Neuhoff, 2003

it must be included as a risk factor in the calculation of the transaction cost of the actors. Therefore the **resulting power** in negotiations does not reflect the optimal investment in power. Any effort to minimise the transaction costs in negotiations must start by reducing the uncertainty, i.e. removing the obstacles in network markets which keep actors from negotiating.

Uncertainty about cost boosts creativity

The cases demonstrate that the need to interconnect in the network markets, which is equal to providing the access to essential facilities, causes negotiations. From their ability to access essential facilities the actors derive **perceived** power. As the network-effects are increased by interconnection all actors can profit from negotiating such an agreement, but it is not clear who profits the most. The asymmetrical power distribution creates different transaction costs for the different actors and is thus a source of uncertainty.

This uncertainty is amplified by the dynamic reactions of the actors. It even seems as if this uncertainty often leads actors to disregard the transaction cost of the negotiation as they are trying to position themselves. But by 'playing' with power, creating 'big' versus 'small' situations, the actors might actually be making their position even less stable. Such forms of behaviour on the actors' part are, however, rational under conditions of uncertainty. The 'big' vs. 'small' effect is also aggravated because companies react differently from individuals in negotiations, as different issues are of different value and importance to them. The regulator, who does not have a complete overview of the particular situation, sometimes makes things even worse. The result of the negotiation is a product of the reaction to the power sources: **resulting** power.

This chapter on the cost level of a negotiation has demonstrated again the close relationship between the source and the dynamics of power. To summarise, it can be stated that transaction costs are a major driving force to negotiate in network markets. If the actors were to accept the differences in perceived power between the actors in their negotiation situation and profit from the complementarities instead of trying to position themselves against each other, they could limit both the uncertainty and the resulting disappointments. **Real** power could hence be applied by the actors together if the high levels of uncertainty were to be reduced. Then the actors could invest more in creating different options for solutions that they can both profit from. In confirmation of **Hypothesis 6**, it can be stated that in this way negotiations can lower transaction cost even in situations of uncertainty.

3.2.3 Strategy – Coopetition bridges the power-gap

Actors explore their possibilities before they start negotiating.[339] In the theoretical chapter it was shown that actors base their decision-making on the experiences they have had in the past and according to the information that is available to them, taking also into account the amount of information they can process. According to this, the key characteristics of the source of power in the strategy dimension are time and information. The dynamics of power are caused by the (individual) decision of the actors to either cooperate or not to cooperate in a negotiation. The overall question for this chapter is: what role does the implementation of strategy in negotiations play in the network markets?

Strategy is characterised by the utilization of information and time

Actors negotiate in order to achieve something which they would not be able to get without a negotiation. Negotiations focus mainly on the price of a certain product (content) and when it will be available (time).[340] These two issues also mirror the major sources of power with – and possibly the difficulties of – the implementation of strategy in negotiations in network markets.

The access to information is fundamental for strategic moves. The actors in network markets have a clear idea of what information they want other operators to allow access to and what they would rather like to 'hide'.[341] To defend their strategic position actors thus supply or withhold information on certain issues in the negotiation situation.[342] The information available (only) to them, together with the know-how of the actors and the ability to process this information, are the basis for their position 'against' the other actor.[343] The following example demonstrates this: in Germany the operators in the telecommunications market do not have to publish their 'business secrets' in the regulatory hearings, but only have to disclose them if the regulator requires them in order to make a decision. It

---

[339] See McKinsey, 2002, for a 'blueprint' on negotiation strategies of an incumbent in a network market.
[340] See RegTP, 2002 and 2003; Monopolkommission, 2004b, for cases in the telecommunications markets. See Schweinsberg, 2002, p. 16, for cases in the rail sector.
[341] See Rochlitz, 2003; Hirschfeld, 2003; Heinrichs, 2003; Scheurle, 2003.
[342] See Die Bahn, 2002b-2005b; Mehr Bahnen, 2003; VATM, 2001 and 2002, for typical position papers which represent this style.
[343] See Hirschfeld, 2003; Heinrichs, 2003; Scheurle, 2003.

is then up to the regulator to decide if this information (or parts of it) will nevertheless be published, so that the competitors can make use of it for their own replies, but cannot take advantage of it in any substantial way. In many hearings so much data is 'blacked out' that the whole file no longer makes sense to those who read it. The market participants question whether all this information really needs to be considered as a business secret.[344] But one way or another, the actors show that by hiding information they can make the market more non-transparent and can thus create more uncertainty: this demonstrates the power of information. Strategic action of this kind limits the ability of the other actors to reply in a meaningful manner.

It is not only the actors but also the regulator that limit the access to information in a negotiation.[345] The regulatory question on the handling of so-called 'mobile virtual network operators' (MVNO) mirrors such a case.[346] There is only a very limited number of mobile operators in each country as frequency spectrum is considered a scarce resource;[347] the operating licenses were therefore either auctioned or awarded on the basis of a beauty contest.[348] Most of the European licenses include a provision that the operators have to allow for so-called service providers to resell minutes of the operators.[349] This was done to create competition even though there are only a few network operators. These service providers depend heavily on the essential facility 'mobile network'. Still they attempt to create extra value on the basis of these networks, to detach their business plans from the interconnection pricing of the network operators. MVNOs, in contrast to service providers, were to be able to create more value by not only depending on one network operator but by being able to offer their services on a variety of (mobile-)networks.[350] The discussion on the revision of German TKG 2004[351] demonstrated that the established network operators tried to fend off the MVNO idea to protect their long term network licenses,[352] by claiming that there is no need for such services, nor for regulation to provide them. This is clearly a strategic move to keep competitors

---

[344] See Schütz, 2005a, pp. 166 ff.
[345] See RegTP, 2004. This is obviously not the case for all circumstances, as for example RegTP publishes a project plan every year.
[346] See Schütz, 2005a, pp. 178 ff., for a definition of MVNO; Bieler, 2005, for the market situation of MVNOs in Germany.
[347] See Kruse, 1992, p. 4, for a discussion on scarcity of frequencies.
[348] See Bornshten/ Schjter, 2003, for an international overview.
[349] See Cullen International, 2005, table 29.
[350] See Tele2, 2005, pp. 15 ff., for an example of such an MVNO.
[351] See Deutscher Bundestag, 2003.
[352] See Spiller, 2004.

and MVNO products away from the markets.[353] As the information on the possible advantages of a MVNO model are not obvious to a regulator, it was fairly easy for the established operators to set up (new) barriers to bar the market entry of new operators so that only they themselves can profit from their essential facility 'network'.[354] It also put the spotlight on another effect of regulation: for regulators to react actors must be able to make themselves heard first. The MVNO issue demonstrates that there is a minimum size and a maximum complexity that a regulator is able and willing to 'protect'.[355] Cases from the rail transport market endorse this finding: it does not seem as if the regulatory authority would help much on the content side.[356] In this sense negotiations between the actors are restricted by the availability of information and the resulting regulatory measures.

Is time phasing a strategic tool? It is not only the content of the issues of a negotiation which has an influence on the negotiation, but also the timing. The so-called 'liberalisation-index'[357] for the rail transport markets in Europe is a good case for demonstrating the 'path dependence'[358] of decision-making in negotiations in network markets. The 'liberalisation-index' illustrates the development of markets and the regulatory measures over time. It shows that the same (strategic) instrument results in a different effect at different times, as actors react differently at different times in a negotiation.[359] Even though actors seem to have access to strategic instruments to achieve certain results in negotiations, they must realise that they only function at and for a certain time.[360] This fact can also be demonstrated with the following example: the formal introduction of a new telecoms law and its implementation are two different steps that provoke different reactions. Most market participants in Germany, for example, have expressed their concern about how the new TKG 2004 will be implemented, as new (legal-) terms have not yet been clearly defined (or tested in court) and the reaction of the market players is not known. The actors sense that this worry can develop to become an obstacle to negotiations in network markets as uncertainty offers the actors on the one hand the possibility of making use of strategies so far unknown, but on the other hand

---

[353] See Wirtschaftsrat, 2004, p. 3.
[354] See Frost&Sullivan, 2004, for an international overview of how this issue is handled differently (based on individual situations) in the different countries.
[355] See Heinrichs, 2004.
[356] See Die Bahn, 2005b, pp. 34 ff.
[357] See Kirchner, 2003.
[358] See Miller, 2003, pp. 170 ff.
[359] See Scheurle, 2003; Neuhoff, 2003; Werner, 2003.
[360] See Rochlitz, 2003; Neuhoff, 2003.

their effect is uncertain as well. This dilemma is mirrored by the following quote: "It can take years to define what is considered to be 'essential'?!"[361] So it is not only up to the actors to decide what effect a strategy has on the time line.

The association of the German telecommunication providers (Arbeitskreis Netzzusammenschaltung und Nummerierung, AKNN) delivers a representative example of time phasing in network markets:[362]

The first phase could be called the founding phase. The original idea of a working-group of all carriers, i.e. new operators and the incumbent, was to develop technical issues of interconnection and numbering together. The association thus laid the groundwork not only for multilateral but also for the bilateral negotiations of the actors in the telecommunications market, as the results of the working groups had to be implemented by more or less all market players seeking to interconnect with others. The AKNN was the first working group of such a kind in the German telecommunications market. Until that time it had only been the incumbent who made such decisions, but for himself (internally), together with the industry. This first time phase, which began in 1996, even before the Telecommunications Act was passed and implemented, was marked by harmony and the main interest was to get the necessary issues resolved, to 'get things done'. These first approaches to the 'self-regulation' of technical issues highlight the fact that the actors see the need to move obstacles (here technical issues) out of the way in order to be able to negotiate meaningfully. The connection between such a self-regulatory approach and regulation is the negotiation which is necessary in both cases, but delivers different results.

The second phase produced the first observable results. Only after the market was fully liberalised did the incumbent and the new operators seem to realise that harmony between them delivered positive results – which was good, but at the same time stoked fears that the other could be gaining more than themselves. After this realisation, so it seems, results were achieved more slowly, partly due to a more bureaucratic handling of the issues: the working group was now set up as a steering committee with many sub working groups that reported to the steering committee on a regular basis. This prepared the ground for more non-transparency between the different group hierarchies, leading to strategic behaviour: the

---

[361] See TK-dialog, 2004.
[362] See Scheurle, 2003, for this specific example. See Poel, 2002, for an overview of all such associations that exist internationally; Kohlstedt, 2003a, gives an overview of the AKNN; Bruce/ Marriottt, 2002, p. 52, present the ACIF as an international example of how such associations can develop to become a support mechanism for all actors.

groups did not work openly against each other, but tried to make sure that they were able to take advantage of the others by, for example, playing for time.

In the third phase the AKNN was used by other market players who were not part of the working groups. The credit for the first results mainly goes to those active individuals and companies responsible for pushing issues forward. Others, mostly actors not involved in the day to day activity of the working groups, noticed that the group could be used either as a scapegoat ('they did not deliver in time') or as a service institution. RegTP, companies and even the courts 'blamed' certain problems on the AKNN. RegTP, for example, tried to keep out of the active discussions at the AKNN, because results for them were a rather mixed blessing: at times it helped that complex technical issues were being resolved, but time frames were never reliable.

In the fourth phase efficient work was no longer possible: a fact which had something to do with the attempt to establish a new kind of professional AKNN. What is known as a common problem in multilateral negotiations with many small and a few large operators then began to manifest itself: actors started blocking issues for no apparent reason, thereby putting issues on the timeline.[363] Decisions were no longer made on the basis of their contents but were more or less decided by the passage of time. Incumbents, more than others, profit from the passing of time and are thus less sensitive to time-mismanagement or delays in the decision-making of such associations. With the original constitution of the group as a burden (especially the unanimity vote) the group was no longer able to move ahead.

In the course of negotiations learning effects are to be expected as some elements seem to be recurrent. Such learning effects could be accelerated through benchmarking.[364] Even if a typical negotiation structure cannot be discovered for every type of negotiation, those actors who negotiate on a regular basis usually do exhibit a comparable behaviour.[365] But the AKNN case also demonstrates the fact that different power settings for negotiations, influenced by the regulator, the consumers and the competitors, exist in different time phases.[366] Most actors have a problem adapting to the

---

[363] See Drahos, 2003, p. 2.
[364] See Bundesministerium für Wirtschaft und Arbeit, 2003, for an example of such a benchmark produced on a regular basis in Germany by the federal ministry of economics.
[365] See Scheurle, 2003.
[366] See Vogelsang, 2001.

speed of change due to the different phases,[367] simply because, for example, they are unable to sustain the enormous investments which replacing their technology would entail. Regulation tries to smooth out such a transition between the different phases, by establishing a gliding path.[368] If actors then implement their strategic instruments in negotiations so as to make use of the time phasing it is likely that they will in due course create more uncertainty.

Résumé: Strategy has a different effect in different time phases. In both network markets, telecommunication and rail transport, essential facilities are the main reason for negotiations and are also at the centre of the respective regulatory regimes. The regulatory developments in these markets are proof of why content and time are the major source of **perceived** power in the strategic dimension. In the beginning of the liberalisation period the main focus in negotiations is on the access and the 'right' price of the essential facilities,[369] to 'get things done' and only later does the demand grow to take account of further issues.[370] The markets analysed here are both based on the same underlying strategic power issues, but today have reached different stages of maturity (in respect to regulation and liberalisation but also in general) in the transformation process.[371] In confirmation of **Hypothesis 7**, it can be stated that in network markets the possibilities of reaching a positive agreement in a negotiation depend heavily on the access to information and the time phase, but these factors are not determined by the actors.

Strategic behaviour causes dynamics

The dynamics of power in the strategy dimension are caused by the decision of the actors, so it was claimed in the theoretical chapter, as to whether they should cooperate or not in negotiations. Is there any reason for a 'small' actor even to try to negotiate at all?[372] To make this decision, can the application of game theory or the use of indicators support the ac-

---

[367] See Schütz, 2005b, p. VI.
[368] See Schmidt/ Winkelhage, 2005, p. 13; Telekom-Control-Kommission, 2005.
[369] See Scheurle, 2003.
[370] See VATM, 2003, pp. 1 ff., as a demonstration that the VATM in Germany seems to have made a leap forward by moving from the 'regulatory-association' to a association that moves forward on (innovative) contents.
[371] See Newbery, 2000; Pelkmans, 2001, p. 455; Bundesregierung, 2004.
[372] See Piepenbrock, 2004. He highlights the fact that conflicts do not disembogue into regulatory cases because of the contents, but because negotiations never take place.

tor in grasping the dynamics of power? The question is if and how dynamics in the network markets cause (resulting) power in negotiations.

How do actors make the decision to cooperate or not to cooperate? In network markets actors compete for the same customers, but need to cooperate on the input side.[373] This is a somewhat paradoxical and asymmetrical starting position for negotiations but one which characterises network markets. As one way of solving this problem actors frequently reduce the complexity of reality. Take the example of the so-called 'competition report' of Deutsche Bahn,[374] where reality is reduced to a simple differentiation between friend and foe, but the complex details of the actual negotiation situations are not considered. The incumbent thereby categorises the market players as those that deliver competing services (foe) and those that complement his own services (friend).[375] The following case illustrates this reduction of complexity in more detail: in 2002 Deutsche Bahn refused to include the timetables of its competitor Connex in their own timetables, but included those of other carriers.[376] DB stated that it was not willing to deliver any services to those with whom they were in competition.[377] The case went to court and one could say that it backfired for DB insofar as it generated bad press for them and they were still finally ordered to list Connex in their timetables. A somewhat similar conflict had arisen a few years earlier in the telecommunications sector, when the new operators asked to be included in the existing phone books of DT. In this constellation the clear language of the telecommunications law ('all must be listed') quickly helped to solve the conflict with regard to the actual listing, but the price of the listing is still an issue of regulatory conflict.[378] The introduction of call-by-call in local networks in Germany also resembles a negotiation, where complexity was reduced early on, which in turn led to subsequent problems.[379] In 1996, before the liberalisation of the public telephony market, the AKNN, then only a small group of carriers including the incumbent, discussed the future of the telecommunications market in Germany. In due course they developed a plan whereby technical adjustments would become necessary in the networks of all the carriers in order to prepare for the introduction of services that were to be liberalised in 1998. In

---

[373] See Engel, 2002b, for a description of how negotiations are run in network markets.
[374] See Die Bahn, 2002b-2005b.
[375] See Die Bahn, 2003.
[376] See Die Bahn, 2004, p. 31.
[377] See Die Bahn, 2003.
[378] See RegTP, 2005b.
[379] See Kirchner, 2002, for background and case; Neumann, 2002.

these discussions the actors came to the conclusion that carrier selection on the local level would not be necessary. With that decision they technically excluded local traffic from the traffic to be routed via new operators. Thus carrier selection on the local level was not introduced as a service in 1998, even though other countries did so at the beginning of the liberalisation of their respective end-user markets. This decision to simplify the market conditions was wrong from an ex post point of view, but it must have been difficult to imagine at that time that the local market would later become an interesting segment in which to invest. This points to the fact that making situations more simple than they are does not necessarily help in the long term, as it is precisely these suppressed details which will need to be resolved later on. This creates dynamics and even more uncertainty.[380]

The way actors behave in negotiation situations could help to answer the question as to whether the decision to cooperate or not is a conscious one. The following example demonstrates that actors tend to make abrupt decisions which are consistent with a sudden change of behaviour: Deutsche Bahn and Connex position themselves as fierce competitors in the German rail transport market. Both have in the past initiated court cases and regulatory cases. But suddenly a change in strategy is announced which seems to imply a u-turn:[381] the former competitors plan to cooperate like partners and will do so without political and judicial confrontation.[382] On the other hand the existence of such a cooperation is publicly denied.[383] This demonstrates that decisions to cooperate or not to cooperate might be made consciously, but have only a very short term effect, as actors do not expect to be able to keep up a non-cooperative style for long.[384] Even within a negotiation there is a shifting back and forth of cooperation and non-cooperation between the actors,[385] caused by the actors themselves.[386] Thus even if it is a conscious decision it is just as unpredictable as an unconscious decision. Or to put it another way, it seems as if the actor would not make a conscious decision to be cooperative or non-cooperative in general but would rather base his reaction on the surrounding factors and his ex-

---

[380] See Scheurle, 2003.
[381] See Rochlitz, 2005.
[382] See Böhmer/ Delhaes, 2004, p. 62; Ott, 2004.
[383] See Krummheuer, 2005b, p. 17. "There is no peace between DB and Connex, because there was no war – only competition and our effort to gain fair access to infrastructure." Own translation from German to English.
[384] See Wünschmann, 2005.
[385] See Heng, 2004a, demonstrating that in the mobile telecommunications sector cooperation between some groups of operators and non-cooperation between others is executed at the same time.
[386] See Hirschfeld, 2003; Röthig, 2002; FAZ, 2002b, p. 67.

perience of the other actors. The number of negotiations that an actor expects to have with the same actors is one of the major reasons for his behaviour in the present negotiation situation.[387] Actors try to minimise their dependency by positioning themselves against the other actors and by moving into strategic positions and executing power.[388] This demonstrates that the 'execution' of cooperation and competition strategies is more a game of positioning ('big' vs. 'small') than a matter of the actual content. It might also be based on a misunderstanding as the actor seems to send signals that he is willing to cooperate, but these are interpreted as signs of competition. In fact Connex, the largest competitor in the German rail transport market, has lowered its pricing by accepting a 'rebate card' ('Bahncard') from its competitor without receiving reimbursement from Deutsche Bahn. DB interpreted this as a confrontation.[389] Even alliances, which are a special kind of cooperation, between incumbents (on an international basis) are not a promising approach.[390] As long as the actors fail to find a way of making their decision-making process more transparent neither of them can define their exact position in the negotiation. To summarise: it cannot be definitively shown that the actors make a conscious decision to cooperate or not to cooperate.[391] Actors do not set a strategy for all negotiations to come. This could be due to the complexity of the negotiation setting. As it is the individual actor who makes the decision on how to proceed with his negotiation partners in his individual situation, negotiations are one way to create more transparency, for example by setting rules of the game together, and thus creating certainty for the actors in their individual situation.[392]

---

[387] See Hirschfeld, 2003.
[388] See Werner, 2002.
[389] See FAZ, 2005b, p. 11.
[390] See Die Bahn, 2002a, p. 21; Heng, 2003.
[391] See Die Bahn, 2004, pp. 23 ff. The incumbent DB describes here the cases that were handled by the EBA in the past year. In describing some of the cases DB tries to show that the reproaches of misuse of power were not conscious moves of Deutsche Bahn. See Die Bahn, 2003, p. 20 and Die Bahn, 2002b, pp. 15 ff., for a complete listing of the EBA cases from 2000-2002. DB differentiates between different classes of misuses and even comes to the conclusion that there are quite a few cases where the incumbent hurt himself the most, to make way for his competitors.
[392] See Die Bahn, 2002b, p. 1, where the incumbent DB points to the fact that transparency is of importance in getting rid of strategic games.

There are situations when actors want to show that they are able to push something through, no matter what.[393] This method of using power is often considered unfair by the actor who is affected by its use and it may lead him to consider revenge.[394] As tit-for-tat is the dominant strategy in negotiations,[395] the attempts to exact revenge lead to further shifts. An example of the 'unfair use of power' is the issue of 'bundling', as in network markets bundling of different services or products carries a negative, i.e. unfair, connotation.[396] Cases from the telecommunications market[397] confirm that the incumbents often try to transfer 'market power' from one market to another by bundling products;[398] the introduction of call-by-call in the local market in Germany is again such an example. Some carriers asked for the introduction of the service and the incumbent campaigned vigorously against its introduction, fearing that its monopolistic market shares would be eroded immediately.[399] The perception of unfairness originates in the combination of issues that at first did not seem to have anything in common, as if the actors had 'made them up'.[400] Another such example is that incumbents tend to threaten to lay off workers if their monopolistic network is to be separated from their services.[401] Even though actors try to gain certainty by threatening they only succeed in creating even more uncertainty. Some actors, though, seem to be able to get away with more non-cooperative behaviour.[402] The term 'fairness' can still be applied to a situation where an actor refrains from applying power to the detriment of a weaker actor.

There is also a positive effect of bundling in negotiations. When negotiating issues, actors prefer to make package deals such as 'I will only cooperate in issue A, if I get issue B' to find a solution at all. This makes sense as the different actors assign different priorities to the issues under discussion. By connecting them, actors try to level the differences out, creating

---

[393] See Die Bahn, 2002a, p. 21. Here it is openly stated that using the market power is seen as a promising strategy. But probably not even DB would consider this to be a fair strategy.
[394] See Werner, 2002.
[395] See Axelrod, 1988.
[396] See Tele2, 2005, pp. 13 ff.
[397] See Vogelsang, 2001.
[398] See Dialog Consult/ VATM, 2004. See Schmidt, 2004, for an opposing opinion.
[399] See Höckels/ Rottenbiller, 2003, for the opinion of the incumbent.
[400] See Essig, 2004.
[401] See Balsen, 2004, p. 9.
[402] See Rochlitz, 2003; Neuhoff, 2003.

more value for the whole package of issues.[403] If an actor limits the negotiations to only one issue, this might lead to a non-cooperative game.

Indicators are implemented to enable one to grasp the complexity of negotiations. Time and content were developed in the theoretical chapter as the major strands of strategy. Strategy is the coordination of the time and content dimensions, i.e. planning and structuring of the next moves and analysing past accomplishments. As strategy is considered a business secret by most actors, there are very few documented sources that describe the implementation of strategy in concrete situations. The only source that mentions the use of game theory for negotiation situations in the broadest sense in the telecommunication sector is the auction of the UMTS frequencies in most European countries.[404] In this case the implementation of game theory was limited to a single decision, i.e. how to make a bid for the frequencies.

Even if not all negotiation situations are documented, it is probably not wrong to suggest that the use of analytically complex models of game theory does not find its way into everyday negotiations.[405] Nevertheless game theoretical applications are used by actors to prepare their negotiation, as a means of trying to grasp complex situations.[406] Conscious strategic decision-making can be illustrated with the following example: DB has internally discussed the different options that arise from the newly competitive situation that they find themselves in. To test these options a benchmark against other markets and countries was developed.[407] This demonstrates that actors develop and plan their actions in advance, think in alternatives and even compare the different network markets in this process.

Game theory is an instrument that helps actors to structure their goals in a negotiation. In combination with experience and know-how this constitutes a decision basis that the actors can make use of.[408] One way of demonstrating the role of experience in setting up such an indicator for decisions is to compare the transition from the first telecoms law to the recently amended version (from 'TKG 1996' to 'TKG 2004') to the transition from no telecoms law to the first ('no TKG' to 'TKG 1996'). The second transition was not as protracted, as most of the actors were by then experienced in their relationship with the legislator and the regulator.[409] This

---

[403] See Hirschfeld, 2003.
[404] See Niemeier, 2002.
[405] See Rochlitz, 2003; Hirschfeld, 2003; Scheurle, 2003.
[406] See TK-dialog, 2004.
[407] See McKinsey, 2002; Die Bahn, 2002a.
[408] See Hirschfeld, 2003.
[409] See Bundesministerium für Wirtschaft und Arbeit, 2003.

also demonstrates that the amount of energy necessary to develop strategy differs between the different time phases.[410] Such an index could be used as a time phase indicator by the actors, giving them an idea of where they are situated at a certain point in time.[411] It could generate some sort of security for the actors even in uncertain environments as it is not necessary to make the strategy itself transparent but the way that strategies are set and assessed by the individual actor.[412] Such a tool enables the actor to find his personal best alternative in his individual negotiation situation. Actors strive for such a stable base but do not seem to be able to find it for themselves, but could need external help to establish such indicators.

The way actors move strategically in negotiations is already heavily influenced by regulation and the 'dance on the fuzzy line' between lawful and unlawful action. The following case demonstrates this: value added service numbers have a bad image in the German market, as they are generally associated with adult content and fraud cases. Obviously some (fraudulent adult content) providers profit from this situation. In conflict cases that deal with billing of the customers, the latter are rather inhibited when it comes to complaining about a service which could be associated with adult content. This has had an effect on the total (telecommunications-) market for value added services, which has grown only slowly in the years since the full liberalisation in Germany. Regulation which cares for a stronger and more transparent separation of adult content and other contents could help to propel the market forwards.[413] The strategic action of smaller companies often seems to be limited to execution rather than negotiations.[414] As the introduction of call-by-call on the local level in Germany stalled, some carriers in Germany just introduced the functionalities of such a service with some technical and innovative turn-around. They were immediately stopped, however, by court orders and RegTP decision, which prevented them from offering the services any longer.[415]

Both cases demonstrate that actors are actually willing to bend the rules, to make use of loopholes, to execute strategic moves,[416] which might then influence a whole market segment. Actors implement power no matter what size the other actor is and tend to use the instruments at hand, even if so doing might entail damage to themselves – just as long as one is not

---

[410] See Tele2, 2005.
[411] See Scheurle, 2003; Armstrong, 1998; Ypsilanti/ Xavier, 1998.
[412] See Hirschfeld, 2003; Neuhoff, 2003.
[413] See Elixmann/ Schäfer/ Schwab, 2004.
[414] See Hirschfeld, 2003.
[415] See Schütz, 2005a, pp. 166 ff.
[416] See Frieden, 2004, for the example of MCI.

hurt as badly as the others.[417] In this respect large operators might have more resources and thus better access to strategies.[418] In the case of the introduction of call-by-call on the local level in Germany the incumbent managed to start a resource-consuming back and forth process of finger pointing. The incumbent argued that RegTP would first need to order the introduction of such a service, whereas RegTP pointed to the industry association AKNN, which in turn pointed to the ministry of economics and RegTP. Tired of this vicious circle, the company 01051 referred the issue to the EU commission. Only in 2002 did the parliament, on its second attempt, adapt the TKG 1996 accordingly. The date for the introduction should have been December 2002. But it was not until December 2002 that the necessary specifications were passed; the economic and legal implementation (the price and the contractual issues) were still not solved. Finally, in April 2003, the services were introduced in two steps: firstly, the so-called call-by-call features and secondly, in July 2003 the pre-selection features. But having too many instruments at one's disposal can result in one's being overpowered; where even more power is needed to control oneself one is not able to direct this power towards achieving certain goals in the negotiation.[419] In such a case even applying the 'right' strategy delivers only a few target oriented results.[420]

The question of why actors in network markets assume such a burden and (still) negotiate, even with those that seem to be more powerful, is of major importance when looking at the strategic aspects and the decision-making process of the actors. The introduction of call-by-call on the local level in Germany points to the different levels of communication between the different actors: the bilateral level, where (some) carriers negotiated directly, and the multilateral level, where a large group of carriers worked on pushing through an industry-wide solution. Then one has the external level and the regulatory involvement, as well as the participation of the public in a hearing and the giving of expert opinions.[421] Finally, there is the political level, which reacted because of pressure from the EU.

This demonstrates, first of all, that actors in network markets negotiate because they are obliged to negotiate. Furthermore, actors are desperately in need of a result and even take into account the possibility of not making a good deal – just as long as it is a deal per se. This 'just do it' way of ne-

---

[417] See Schmidt, 2005a, p. 14, demonstrating that it seems to be common knowledge that the actors play foul.
[418] See Hirschfeld, 2003; Heinrichs, 2003.
[419] See Heinrichs, 2004; Scheurle, 2003.
[420] See Michalowitz, 2004, pp. 269 ff.
[421] See Schwerhoff, 2002.

gotiating seems to be widely represented in network markets.[422] Another reason is that the actors do not know in advance if it will be a difficult negotiation situation, where they will be less powerful than the other actor.[423]

Actors do not have to make a decision once and for all regarding the need to cooperate or not as it is possible to cooperate in (some) negotiations, but to compete at the same time in other segments. In the case of 'call-by-call on the local level', most of the actors at first chose the non-cooperative behavioural option as a means of trying to succeed with their own position. Those carriers that have multiple interests, i.e. both in the long-distance and in local market, kept quiet. It is interesting to see ex post that all the actors involved (including RegTP) praised the fact that most of the technical work had been completed surprisingly quickly – in cooperation. Another example demonstrates that actors in network markets have an interest in cooperating even though they might be competitors: two telecommunications carriers have signed a Service Level Agreement (SLA) for the provision of a leased line service, where each actor supplies the other with one half of the whole line. If one actor breaks the agreement and delivers a quality level that is less than was agreed in the SLA, the other actor might be tempted to intentionally lower the quality on his half of the line. If he chooses instead to maintain the quality as promised and helps improve the quality of the other half of the line, he might be better off in the long run, as he can expect the other actor to help in case of malfunction on his side as well. As there is a constant shift between cooperation and non-cooperation, eventually an arrangement between cooperation and competition might be the result: 'coopetition'.[424]

Résumé: The non-transparent decision to cooperate or not to cooperate aggravates the asymmetries in negotiations. The cases show that the actors explore the instruments at hand, but do not necessarily take into account what they might possibly destroy by behaving in that way. In the case of call-by-call in the local area, some actors were able to profit from a first mover advantage by not moving at all. With less use of strategic power the situation could possibly have been resolved more quickly. In the long run holding on to a power position does not seem to pay off. The non-transparent decision to cooperate or not to cooperate aggravates the asymmetries between the actors and causes dynamics in negotiations: **resulting power**.

---

[422] See Heinrichs, 2004; Scheurle, 2003.
[423] See Rochlitz, 2003; Hirschfeld, 2003.
[424] See Ritter, 2004, p. 125. She states that this applies to the situation in the German telecommunications market.

The use of strategy seems to be a synonym for turning competitors against each other, to make the best out of the negotiation situation (only) for oneself. Everything that is available will be used to build up power, even though it does not seem clear for the actors what they can or want to achieve with that amount of power. In confirmation of **Hypothesis 8**, it can be stated that game theory as such is not implemented to a large extent, but can support negotiations as an indicator which enables one to grasp the dynamics of power in negotiations. Still the main goal of most strategies, as the cases demonstrate, is to cause uncertainty for the other actor and more certainty for oneself. This reminds one of the arms race during the cold war, where all involved invested heavily in a power set up, but none knew what good (or bad) it could do.[425] Thus power and the use of strategy are closely connected.

Strategy supports the individual in choosing the best approach to a negotiation

The cases show that power is not evenly distributed but accessible, albeit in different amounts, to all actors in network markets. The sources of power in a strategic context are timing and content. Actors have an image of the effects of these sources of power in negotiations, which is **perceived power**, but are uncertain about the results they can achieve by implementing them. It seems as if some actors can profit, but only in the short term, from such a situation of uncertainty.

The use of the instruments causes shifts of power in the negotiations, making the results for which the actors strive uncertain. In trying to balance these shifts, actors counterbalance with their own shifting between cooperation and non-cooperation. Very rarely and more by chance than otherwise, the actors achieve exactly what they were striving for. In most cases the dynamics lead to power-negotiation **results** of inferior quality – for all actors involved. Strategy would then partly be based on a conscious decision-making process, but actors would not be conscious of the results and possible restrictions that might arise for all involved from the use of such a strategy.

The asymmetries 'produced' by the actors by their 'decision' to exhibit non-cooperative or cooperative behaviour in their individual negotiation situation can be overcome with the help of transparent negotiations. In confirmation of **Hypothesis 9**, it can be stated that the way strategy is commonly applied seems at first sight not to relieve the actors of uncertainty, but instead to create more of it. Realising the source and making use

---

[425] See Schelling, 1957.

of the dynamics of power, i.e. aligning the strategic powers, is what strategy in negotiations is all about. The **real power** in the strategy dimension is the common production of options that are of advantage for all actors in the due course of negotiations.

3.3 Relevance of the findings for negotiations

In this chapter of the study the theory on the asymmetrical distribution of power was applied to case studies. For each dimension of the Power-Matrix the question as to whether power is a relevant issue in negotiations in network markets, such as the telecommunications- and the rail transport markets, was discussed. The analysis demonstrated that both the source and the dynamics of power are an integral part of negotiations in both markets. In confirmation of **Hypothesis 10**, it can be stated that the characteristics of power and negotiation are interrelated: the Power-Matrix and the analysis of negotiations in the three dimensions framework, cost and strategy makes evident the fact that power, asymmetry and negotiations are linked to each other.

There are not only similarities between the network markets under consideration with respect to the economics of negotiations, but these network markets are also closely connected and in parts even depend on each other.[426] Looking at more than one market (-segment) at a time therefore offers the opportunity to compare similar developments, but at different stages on their time-path[427] – one is thus able to develop improvements regarding the handling of power in negotiations. The above analysis of the rail transport and telecommunications sectors has demonstrated that the degree to which negotiations are a product of choices made by the actors and regulators in each separate market depends on how far that market has developed. The comparison of rail and telecommunications could even be considered to be both anti-thesis and pro-thesis (fast moving, global market against slow moving, rather national market), a fact which broadens the

---

[426] Take the example of the German market, where Arcor, the second largest telecommunications operator, was originally founded on the basis of the telecommunications network from Deutsche Bahn. The existence of telecommunications networks in the rail sector is based on the need for high quality information networks. For more details on these networks see Matthies/ Schinkel, 1991.

[427] See Miller, 2003, p. 170; Die Bahn, 2004, explicitly puts the different network-markets telecommunication and rail into a context where one market can learn from the lessons of the other.

perspective on the possible use of power in negotiations and exhibits learning effects as to how to best adapt to the usage of power (over time) even in other (network) markets.[428] It is in this light that in Germany a 'super-regulator' for infrastructure is being established to cover rail transport, energy, telecommunications and post.[429]

---

[428] For a critical view see McKinsey, 2002, p. 2.
[429] See Rochlitz/ Winkler, 2005; Neuscheler, 2005, p. 45; Bötsch, 2005, p. 13; Schütz, 2005c; Bundesrat, 2005; FAZ, 2005f, p. 14.

4 Alternative Dispute Resolution enables efficient negotiations

It would be ambitious to assume that negotiations in markets where power is distributed asymmetrically are an easy task. Nevertheless there could be something like an optimal negotiation structure just as there are typical 'mistakes' and reasons for failure in negotiations in special markets. The Power-Matrix is of help in establishing such a structure.

This chapter is laid out as follows: firstly, the results from the theory and the cases above are summarised to define the key focus of this chapter. What are efficient negotiations and what goes wrong in so many cases and why? The Harvard Negotiation Project[1] is introduced as a mechanism on the basis of which Alternative Dispute Resolution[2] (ADR) instruments can be set up to offer support in negotiations. Secondly, the three steps by which these promising alternatives ought to be implemented, co-ordinated and initiated will be discussed. Finally, a short conclusion sums up the study with a concrete recommendation on how to actually effect a change in the system of negotiations in network markets by establishing a basis for the promotion of efficient support processes.

4.1 Alternatives to 'classic' negotiations

The measure for 'better' negotiations is Pareto efficiency: in this individual situation there are no better solutions presently available, but this does imply that under different circumstances better solutions could be achieved.[3] Pareto efficiency does not describe a state where the actors are

---

[1] The authors of 'Getting to Yes' (Fisher/ Ury/ Patton, 1991) are also the founders of the so-called 'Harvard Negotiation Project' which stands for interest-based/ principled negotiations. I have chosen to use the term 'Harvard Negotiation Project' in this study to reflect the whole idea and not just to focus on one aspect of it.
[2] See Bruce/ Marriott, 2002.
[3] Better solutions might theoretically be available, but are, for example, not pragmatic.

'symmetrically' happier, but it is sufficient that both have experienced nett gains compared to the initial situation.[4] On the passage to efficiency in negotiations there are, as was shown above, problems and pitfalls. To counter these problems it could be necessary to find an alternative to the way negotiations are run today – 'classic' negotiations – or even to develop an alternative to negotiations altogether. The cornerstone of such a model needs to be a commitment to tackling the key problems in negotiations. Obviously, another solution could be to prohibit conflict.[5] Such a solution would probably only lead to conflicts at another level or time and would most likely even accelerate the original conflict. Since the prohibition of conflict is not a realistic possibility it will here not be considered further.

Firstly, the cases described above will be analysed and the problems will be structured so as to enable the development of possible solutions. Secondly, the Harvard Negotiation Project is introduced as a behavioural scheme, which could, merged with the Power-Matrix, deliver a promising basis for alternatives to negotiations. Thirdly, alternatives to 'classic' negotiations will be analysed that can handle complex negotiation situations.

4.1.1 Is power a problem for negotiations in network markets?

Negotiations in network markets, despite all the possible problems, take place every day. To make them more efficient it is necessary to analyse the problems. Since the key characteristics of power in negotiations could also be the key problems of negotiations in network markets, this chapter first structures the problems along the Power-Matrix and considers whether power is the main or just an underlying problem in negotiations.

The **property rights** distribution in network markets is heavily influenced by regulation. This is deemed necessary as the process of liberalisation by regulation is thought to be the Achilles heel of the development of network markets.[6] But regulation also causes problems in network markets because regulators are not reckoned to be very innovative;[7] they cannot even keep up with the development of new technologies.[8] It is therefore difficult for them to describe the best route to competition.[9] Yet regulators still try to achieve a situation where power is distributed equally between the actors. The common understanding is that an equal distribution of

---

[4] See Breidenbach, 1995, pp. 76 ff.
[5] See Külp, 1965, p. 97.
[6] See Waesche, 2003, p. 2.
[7] See Gifford, 2003, p. 477.
[8] See Schmidt, 2005b, p. 19.
[9] See Peltzman/ Winston, 2000.

power tends to result in more efficient negotiations.[10] Thus it seems that in situations where there is a 'balance of power'[11] the number and the intensity of shifts could be reduced or eliminated. That way the actors would be certain of their exact (power-)position within the negotiation as everyone would be in the same position. Such a balance of power does not mean that there is no power in the game, though;[12] it is only that there is no further value in using it. This could, obviously, be different in a subsequent round of negotiations. This is why symmetry in negotiations cannot be a (regulatory) goal in itself, as it does not describe a steady state. The veil of uncertainty[13] about one's (starting) power-position in a negotiation could be a temptation for the actors to use power, in order to avoid being the only ones not using it. For most negotiation situations this means that to have 'weapons' of power at one's disposal (mis-)leads the actors as to the power they **perceive** they have.

To counter this problem any alternative concept must therefore address the source of uncertainty. Uncertainty as a problem could be tackled by making negotiations a more central part of regulatory regimes.[14] Negotiations are already a part of the regulatory obligation in most network market regimes today, but the way in which they should be run is not explicitly mentioned. This demonstrates that it could be necessary to set up a framework which allows one to create a process of negotiations which is more adapted to the dynamic environment.[15] Such a framework would provide the rules of the game, but not a regulator who would interfere with issues that the actors could resolve themselves. It would also allow the actors to identify the right level upon which to act.[16]

Essential facilities are the core characteristic of network markets and are the reason for 'big' vs. 'small' negotiation situations. Such situations are ripe for the use and abuse of power.[17] Uncertainty about one's own exact 'size' makes it impossible for the actors to tune in their instruments of power so as to create a positive result in a negotiation by anything other than mere chance. Additional power in this respect might not necessarily be better and could actually be worse.[18] The reason for this could be that

---

[10] See Zartman/ Rubin, 2000, p. 15.
[11] Harsayni, 1956, p. 145.
[12] See Emerson, 1962, p. 33.
[13] See Rawls, 1971, pp. 136 ff. The term 'veil of uncertainty' is an adaptation of the original term 'veil of ignorance'.
[14] See Kohlstedt, 2003b.
[15] See Beardsley/ Enriquez/ Garcia, 2004, pp. 48 ff.
[16] See Berg/ Foreman, 1996, pp. 650 ff.; Lapuerta/ Tye, 1999, p. 138.
[17] See Seeley/ Gardner/ Thompson, 2002, p. 3.
[18] See Maoz, 1989, p. 239.

the actor with seemingly more power might have too complex a (hierarchical) decision-making structure, which prevents him from applying his power efficiently.[19] Additionally there is a good chance of weaker actors getting stronger actors involved in negotiations and then borrowing power (for the moment) from them. 'small' actors even get support as a minority (e.g. in the case of asymmetrical regulation); altogether they are able to make better use of their resources. The constant fear of the swinging pendulum of power changing the power distribution between the actors keeps the actors negotiating. Both actors know that in a court case, for example, there will be a winner and a loser; but one cannot predict who will be the winner and who will be the loser. In a symmetrical situation, no actor would be able to move if he were trying to preserve the symmetry. To put it bluntly: 'balance is death'.[20] In that respect the actors are more cooperative in an asymmetrical situation, insofar as they are making room for a creative process to solve the problems.[21] Thus power asymmetries can be put to use to create efficient negotiation results, e.g. **resulting** power, and in this sense weakness can even be strength.[22]

In the light of this discussion it seems as if it is not asymmetry which is a problem for negotiations, but the use of power to battle against asymmetry.[23] To counter this problem any alternative concept must hence deliver the basis for a creative negotiation process. Instead of using power against each other, for example by making use of legal loopholes, the actors can make better use of their powers by finding an efficient solution for all involved.[24]

Timing and the **access to information** constitute the basis for strategic action, i.e. the actor's decision to execute a cooperative or a non-cooperative negotiation style. There seems to be a constant temptation for actors to operate strategically in order to maintain or to gain the more powerful position, even though this does not necessarily correlate with a better result in the negotiation. Due to the complexity of a relationship and the resulting negotiations hardly any situation of complete asymmetry exists. Complete asymmetry may be described as a situation where A influences B, but B does not influence A at all. Incomplete asymmetry would be a situation where both actors have some kind of influence on each

---

[19] See Zartman/ Rubin, 2000, pp. 263 ff.
[20] See Pascale et al., 2002, p. 18. According to them this hypothesis is supported by events in nature.
[21] See Zartman/ Rubin, 2000, p. 30.
[22] See Schelling, 1956, p. 306, for this expression.
[23] See Zartman/ Rubin, 2000, p. 257.
[24] See Berg/ Foreman, 1996, p. 651; Kohlstedt, 2003b.

other.²⁵ In situations of incomplete asymmetry, i.e. where the actors are mutually dependent, negotiations make sense, because of uncertainty as to the (exact) extent of the other actor's power. A cooperative approach to stopping conflict, maybe even before it breaks out, is a long term route to successful cooperation and thus the way to **real** power.

To counter strategic activity which is mutually detrimental any alternative concept must create a situation where the actors can identify common points of interests by taking a peek behind the other's position. In that way everyone would profit from a transparent set of strategies, which would make the use of power issue specific, and therefore leave room for the individual to choose his best alternative in the negotiation situation.²⁶

To summarise: it is likely that the key characteristics of power in negotiations also pose the key problems in negotiations. What makes negotiations even more difficult, though, is that different instruments of power are used in negotiations, which in turn create different effects in the different phases of the negotiation, but no one knows when and how.²⁷ Thus asymmetry must be seen as the core concept of power. It can be stated that asymmetry as both the source for power and cause for the dynamics, makes power a relevant issue in negotiations. Handled correctly, power asymmetries help to set up an efficient process in negotiations.²⁸ The Power-Matrix with its three dimensions helps to structure the possible problems with power in negotiations as a first step to designing the necessary cures.²⁹

4.1.2 Merging the Power-Matrix with guidelines on how to run negotiations – The Harvard Negotiation Project

This chapter merges the Power-Matrix and the so-called Harvard Negotiation Concept in order to design a supporting process for asymmetrical power-negotiations. The concept of the Harvard Negotiation Project is a very simple and intuitive process description.³⁰ As a "well structured common knowledge"³¹ it seems to be an easy, widely accepted guide,³² de-

---

[25] See Simon, 1957, p. 66.
[26] See Zartman/ Rubin, 2000, p. 261.
[27] See Kirchner, 2003 and 2004.
[28] See Zartman/ Rubin, 2000, p. 3.
[29] See Lewicki et al., 1994, p. 143; McKinsey, 2002, for a similar structure.
[30] See Wagner, 1995, p. 9, addressing the problem that the focus in negotiation research is mostly on the output rather than on the process of the negotiation.
[31] See Hättenschwiler, 2003.
[32] See Alexander, 1998, p. 52.

livering the basis for successful negotiations.[33] What is the difference between a classic and a Harvard Negotiation Project approach to negotiations? In the traditional 'positional bargaining' (or distributive/ zero-sum bargaining)[34] actors each assume a position and from then on make concessions in order to reach a compromise. The Harvard Negotiation Project is based on so-called 'principled negotiations', also called interest based negotiations,[35] and the opposite of 'positional bargaining'.[36] Principled negotiations are said to provide more satisfying results for the actors. Other than in classic, i.e. position based negotiations, actors do not limit themselves to certain legal or technical positions but also go into their needs and interests.[37] This makes the concept flexible to dynamic changes even within negotiations.[38] It is in this sense that the Harvard Negotiation Project could provide guidelines which would enable actors to operate along basic principles even in network markets and in situations where power is distributed asymmetrically between the actors. Here the design of the Harvard Negotiation Project will be tested to ascertain whether it can provide general (economic) guidelines for (difficult) negotiations.

The Harvard Negotiation Project is made up of several phases that grasp and describe the different stages of the negotiation process. The exact number of phases (between four and seven) depends on the 'translation' of the core concept.[39] The original four phases are as follows:[40]

1. Separate the People from the Problem.
2. Focus on Interests, not Positions.
3. Invent Options for Mutual Gain.
4. Insist on Using Objective Criteria.

The authors of the Harvard Negotiation Project describe a basic concept of how to run negotiations, which can and must be adapted to the individual situation. Thus it is only natural that the number of phases varies ac-

---

[33] See Barendrecht, 2003.
[34] See Bruce/ Marriott, 2002, p. 12.
[35] See Alexander, 1998, pp. 50 ff.
[36] See Funken, 2002, p. 5, pointing to the fact that there are situations where neither a principled negotiation nor a good choice are possible. Those are situations where single issues are negotiated and the actors will never again negotiate. Such situations are highly unlikely in the rather small but complex network markets.
[37] See Alexander, 1998, pp. 50-52.
[38] See Alexander, 1998, p. 3.
[39] See Egger, 2002. The German training version, for example, has six steps instead of four.
[40] See Fisher/ Ury/ Patton, 1991.

cording to the individual interpretation. In any case these phases, no matter what their exact number might be, deliver a structure along which the actors can run negotiations efficiently. No matter how many 'little' intervening steps are needed to reach the respective levels, the main phases of the negotiation process can better be defined by the **achievements at the end of each phase**,[41] here paraphrased with the following imperatives: **'take off the chains'**, **'create added value'** and **'cut the cake'**.[42] There will follow a discussion as to how both the concepts of the Harvard Negotiation Project and the Power-Matrix can be merged on the basis of these three core common steps for negotiations in markets with an asymmetrical distribution of power.

'Take off the chains'

The Harvard Negotiation Project suggests that actors separate the 'people-problems' from the actual issues that have instigated the negotiation. On the one hand such a separation makes it possible for the individual actor to find out more about the interests informing the positions of the other actors.[43] What does this mean? Behind the problems which come up in negotiations lie interests based on values,[44] possibly on the instructions or directives given by someone within a company that are to be obeyed by one or more actors. Setting up and protecting such a position is costly and consumes a great deal of energy. Still it seems at first that less effort is needed for an actor to hide behind a position which has been set up in this way than it takes to go into details. Often, therefore, a position consists merely of a 'yes' or a 'no', with no explanation given as to what this position is based upon. The Harvard Negotiation Project argues that asking oneself what the other actor might think, by trying to put oneself in his position, allows the other actors to explore the interests better. On the other hand it might be difficult to draw a meaningful and theoretically stable clear-cut distinction between interests and positions.[45] Even the theoretical distinction, though, helps to set up a more stable structure, i.e. working relationship, in the negotiation situation.

---

[41] See Hättenschwiler, 2003, pp. 9 ff.
[42] See Hauser, 2002, pp. 12 ff. He distinguishes between three different steps: the constructive-step (here called: take off the chains), the integrative-step (here called: create added value) and the distributive-step (here called: cut the cake). See Barendrecht, 2003, pp. 21 ff.; Seeley/ Gardner/ Thompson, 2002, p. 17.
[43] See Funken, 2002, p. 4. She insists, in response to critics of the concept, that actors should not focus solely on interests, though.
[44] See Barendrecht, 2003, pp. 21 ff.
[45] See Funken, 2002.

The Power-Matrix in the **framework** dimension could support this part of the process by providing the actors with access to the different levels of a negotiation. Together both the Harvard Negotiation Concept and the Power-Matrix can help the actor in his negotiation situation by creating 'transparency'. But what does transparency signify in this context? Transparency allows the different actors to understand the matter which is being negotiated in the same way and at the same time (process, content and time). In a transparent negotiation situation all actors are sure that they are talking about the same issues. For transparency to be established it is not necessary to support the perspective of every actor, but instead it is necessary to accept the other's perspective as it is. If an issue is unclear or an 'old' issue is in the way, this will be clarified before the (new) contents are further discussed. These are the process issues of a negotiation. If the process and the contents of the negotiation are not separated from each other there is a danger that the negotiation will never get beyond the process issues, as they are already sufficient reason for an escalation.[46] Thus transparency helps the actors to structure their individual negotiation approach in order to make rational decisions[47] and to design negotiations which are substantially more efficient.

No one situation is appropriate for distributional bargaining. Some negotiations will always remain zero-sum games.[48] The Harvard Negotiation Project tackles this problem by dividing up the different issues within a negotiation into those where an agreement can be made and those where the actors 'agree to disagree'[49]. This is consistent with the need to separate content and personal issues and also implies the need for a time-phasing of different issues. The actors cannot handle everything at the same time without losing their grip on the negotiation. That means it can be necessary to split up a negotiation into several more manageable parts which can be handled piece by piece. At the same time one is not to forget that these pieces also influence each other. Furthermore, the quality of a (subsequent) agreement in a negotiation will be determined by the way the conflict resulting from a previous disagreement was resolved.[50]

Therefore even in negotiations which are perceived to be hopeless, the actors expect to identify win-win situations.[51] Translated into the language

---

[46] See Becker-Beck/ Beck/ Eberhardt, 1998, p. 108.
[47] See Kahneman/ Tversky, 1991, p. 4.
[48] See Funken, 2002.
[49] See Aumann, 1976. This expression is indebted to the title of his paper: 'Agreeing to Disagree'.
[50] See Esteban/ Sakovics, 2003.
[51] See Hättenschwiler, 2003, pp. 44 ff.

of the Power-Matrix, win-win situations in negotiations can be realised by establishing a high degree of transparency, thus 'unlimiting' the subject, i.e. 'taking off the chains', and getting onto the relevant level of the negotiation. Contrary to common knowledge, smaller actors may be at an advantage when starting such a process, as they are better able to put positions aside and invest in the negotiation process, rather than in power positions. Size is not the decisive factor in making negotiations successful, though: actors admit that some experience is needed in order to understand and to implement the Harvard Negotiation Project in their day to day business.[52] What is even worse is that it might be difficult to convince those who have never worked with the Harvard Negotiation Project to make use of it. When perceptual and cognitive biases[53] make it impossible for the actors to 'talk' to each other, someone such as a regulator might be of help. By implementing structural elements in a negotiation the actors are then able to 'take off the chains' that keep most actors from expanding the 'cake'.[54]

At the end of phase one in a negotiation the actors are hence ready to pose a so-called 'creative' question. This means the actors are willing to call it quits with regard to the previous causes of problems in the negotiation and are ready to start developing a solution targeted at building up their future relationship. In due course the actors learn to make use of their common interests. This corresponds to the framework level of the Power-Matrix.

'Create added value'

The Harvard Negotiation Project aims to create an efficient negotiation result even when resources are limited. By making use of the instruments of the Harvard Negotiation Project the actors develop options together that create value for both even if one actor has better access to an essential facility than another. Starting from the basis of the lowest common denominator it is incumbent on the actors to develop new 'common solutions' together.

The **cost** dimension of the Power-Matrix can support the actors in making best use of their scarce resources. The concept of power for the moment makes it obvious to the actors that symmetrical and asymmetrical settings might be very close to each other, at least on the timeline; uncertainty prevails as long as nothing is done to actively create certainty for the nego-

---

[52] See Hättenschwiler, 2003, p. 60.
[53] See Funken, 2002.
[54] See ERG, 2003a, for such structural elements.

tiation situation. As the investment in a power position might be the reason that negotiations are slowed down, the Harvard Negotiation Project tries to work as a conflict prevention mechanism, stopping conflict before it causes cost.[55] The Harvard Negotiation Project approach is to invest in order to run the actual negotiation process well instead of trying to achieve either asymmetry or symmetry. To reduce the complexity of the negotiation situation, 'a little bit of story telling'[56] about the alternatives that are or are not available can support this process. By pointing to the opportunities that exist for all, by accepting that starting from a new position and looking forward represents an opportunity for all involved, value can be created. The Harvard Negotiation Project aims at considering issues from different perspectives, and at making the problem of the other actor their own problem. The actors can then develop new options together, without being restricted by any framework save that which they have thus developed together. Still, the creativity demanded by the process of developing new options is often limited by the behaviour of the actors insofar as they move too fast, without regard to the input of the other actors.

To translate this into the language of the Power-Matrix, creating a common basis to developing solutions in a win-win process will "maximally accommodate the interests of both actors."[57] The process and the relationship between the actors is more important than the resources that are available to create common value.[58] These solutions are then Pareto superior if not even Pareto optimal. Even though the cost aspect is not 'actively' taken up in the Harvard Negotiation Project, the exchange of information and options between the actors en route to finding the 'right' conditions for the negotiation is not meant to be costless. The process which leads to the development of new and innovative options allows for mutual gain, even though the negotiation process itself causes transaction costs.

Some argue that all grave conflicts are the consequence of disputes about fairness and thus essentially value issues.[59] This would be to support the need for the inclusion of a value discussion or at least an acceptance that values form the basis for conflicts in a negotiation process. What most

---

[55] See Bruce/ Marriott, 2002, p. 12.
[56] See Poitras/ Bowen/ Byrne, 2003.
[57] Barendrecht, 2003, p. 26.
[58] See Lavie, 2003, p. 24.
[59] See Montada, 2000, pp. 38 ff. He claims that it is a deficiency of the Harvard Negotiation Project that it does not discuss value issues. Nevertheless this objection highlights that any concept for efficient negotiations must be based on values, but it obviously depends on the individual situation in the negotiation how that is handled.

negotiation concepts lack, though, are the tools to help the actors articulate and respect these values and thereby to make them a creative part of the negotiation.[60] Evaluating the options and hence making value judgments too early in the process can jeopardize the negotiation as a whole, though.[61] For that reason it is necessary to develop options first and then, only at a later stage, to make a decision as to which options one shall develop further. Such a transparent process, which separates the collection and the evaluation of the options, can help to avoid the problems.[62] The creation of new options that are of help to all actors can be achieved by including or excluding different issues so as to create a larger 'package' (possibly on different levels).[63] The passing of time alone, the progression into another time phase, might also have an effect on the size of the 'cake'; if the horizon is stretched, more solutions can be accommodated.

The difficult part for the actors in a negotiation is to prepare the common ground away from the yes/ no extremes, to make solutions available that are efficient for all involved. The investment in building relationships seems to be a worthwhile investment, in particular – perhaps – for the actor who is perceived to be 'weak'.[64] The Harvard Negotiation Project supports actors in limiting uncertainty and reducing the complexity of negotiation situations. Together with the Power-Matrix such situations can thus be turned into constructive negotiation situations instead of a destructive zero- or negative sum game.

At the end of phase two the actors have therefore developed a choice of possible options that are relevant and capable of being realised. The Harvard Negotiation Project 'respects' existing obstacles as they are perceived by the actors or others, but tries to overcome them by letting the actors work on the problem together, discussing the details of alternative options and thus increasing the chances for a win-win solution, instead of working against each other and 'cultivating' roadblocks. In this process the actors establish a firmly grounded relationship, which helps to reduce uncertainty and creates extra value even in the long run. This corresponds to the cost level of the Power-Matrix.

---

[60] See Montada, 2000, p. 37.
[61] See Becker-Beck/ Beck/ Eberhardt, 1998, p. 111.
[62] See Montada, 2000, p. 60.
[63] See Kals, 2003, p. 6.
[64] See Zartman/ Rubin, 2000, p. 258.

'Cut the cake'

The Harvard Negotiation Project aims to support the actors in discovering options which can be turned into possible solutions by adjusting them to meet the needs of the actors. To support this process, the Harvard Negotiation Project has created a scale of its own, the so-called 'Best Alternative to Negotiated Agreement' (BATNA).[65] In the language of economics, the BATNA is the 'reservation price' and reflects the opportunity costs of the actors.[66] The actors hence develop a choice of alternatives through 'reframing', in this case by combining the common benchmark and their individual BATNA. The result is the zone of possible agreement (ZOPA)[67] within their BATNA. But there are practical limits to the use of objective criteria.[68]

This is where the Power-Matrix might be of help. On the **strategy** dimension it delivers support for the actors in setting up the objective criteria which enable them to derive their individual strategic decisions in the negotiation situation, thus allowing both to get the most out of the negotiation situation. To be able to do so, it is in the interest of all actors to establish rules that are transparent and fair right from the start.[69] Even the decision on how to 'cut the cake' is based on criteria which allow all actors to make an individual decision which is in their own best interests. By and large, it would seem, the actors know which rules and what kind of benchmark suits them best in their individual negotiation situation. Objective criteria and scales such as game theory provide them with help in setting up support mechanisms, which even absorb some of the situation's dynamic content. This part of the negotiation is the strategic heart of the decision whereby the individual actor exercises his judgment on how to share, to cooperate or to not cooperate with the other actors.

At the end of phase three the negotiation therefore concludes with a deal. The overall goal is to move the Pareto border and in this way to be able to cut the bigger 'cake'. The concept of the Harvard Negotiation Project supports actors in finding or developing transparent criteria which can be accepted by all. Only then will each actor be able to choose their best alternative, whilst both can share the fruits of the negotiation agreement. This corresponds to the strategy level of the Power-Matrix.

Figure 4 gives an overview of the merging of the Power-Matrix and the Harvard Negotiation Project.

---

[65] See Fisher/ Ury/ Patton, 1991, pp. 97 ff.
[66] See Ayres/ Nalebuff, 1997, p. 16.
[67] See Blount, 2002.
[68] See Funken, 2002.
[69] See Buchanan, 1966, p. 30.

	Framework	**Cost**	**Strategy**
Power-Matrix Source **and** dynamics of power in three dimensions	Property rights distribution **and** access to different levels of the negotiation	Essential facilities **and** uncertainty and non-transparency	Time and information **and** cooperation and non-cooperation
Harvard Negotiation Project Alternative Dispute Resolution mechanisms as a basis for setting up a supporting process	Separate individual from common issues to 'take off the chains'	Develop new, innovative options to 'create added value'	Make decision to cooperate or to not cooperate, i.e. 'cut the cake'

Figure 4: Merging of Power-Matrix and Harvard Negotiation Project
Source: own design

4.1.3 Alternative Dispute Resolution mechanisms support negotiations

> *"Most people will not associate the legal system with cooperation."*
>
> *Maurits Barendrecht*[70]

At the beginning of a negotiation actors make the decision to either exit, escalate or to proceed.[71] 'Voting with one's feet' is an 'exit' solution and the fallback alternative for every actor in every situation. If actors think they are trapped in the current negotiation they can 'escalate' the situation, i.e. involve more or different actors (for example those on a different level of the hierarchy) and/ or change the speed of the negotiation. On the one hand such an approach raises the costs for all involved, as 'shifts' of the situation are caused intentionally, which can generate further uncertainty. On the other hand an escalation makes room in which to discover different solutions. When the actors escalate or proceed with the negotiation they make an additional decision either to 'delegate' the negotiation to someone else, i.e. to use external support, or to proceed alone. The negotiation tools

---

[70] Barendrecht, 2003, p. 62.
[71] This paragraph builds on Hauser, 2002, pp. 185 ff.

can thus be divided into those where the actors negotiate unaided and those where an external third party supports the process.[72]

Below, the nature and benefits of Alternative Dispute Resolution mechanisms will first be analysed. Then, with the help of a short case study the performance of ADR mechanisms in complex negotiation situations will be demonstrated. Last, but not least, there will be a discussion as to whether the choice of the different ADR mechanisms could deliver a tool-box for efficient power-negotiations.

ADR – the degree of intervention makes the difference

Alternative Dispute Resolution mechanisms were originally based on negotiations between the actors. The term 'alternative' indicates that these instruments were distinct from litigation. The most concise definition of ADR would hence be to speak of 'alternatives to litigation',[73] as litigants allow a third party to decide their case. The court-option is the typical win-lose situation, a very formalised and mostly time consuming process. Actors do not expect much fairness from legal proceedings in court,[74] but where, for example, decisions are made on matters of principle, the court is seen as the best possible venue available. At the very least, courts and their legal decisions thus offer a point of reference in negotiations, even though the obstacles to becoming entitled to make use of legal justice are seen as too high.[75] Nevertheless actors are constantly searching for more (instant) certainty and a higher degree of cooperation.[76]

It has been sufficiently demonstrated that it is the degree of intervention which makes the difference between those negotiation mechanisms in which third party support is applied.[77] In other words the difference lies in the degree of power that the third party has to make decisions for the actors, or how he assists the negotiation process in other ways.[78] For the actors the difficulty lies in finding the right mechanism and amount of intervention for their individual situation. They sense that there are different

---

[72] See Lewicki et al., 1994, pp. 349 ff., for an overview of third party interventions and of which type is appropriate at a particular time. Raiffa, 1982, p. 22, distinguishes between facilitator, mediator, arbitrator and rules for the manipulator.
[73] See Bruce/ Marriott, 2002, pp. 7 ff.
[74] See Mähler/ Mähler, 2000, p. 13.
[75] See Commission of the European Communities, 2002a; Racine/ Winkler, 2003.
[76] See PriceWaterhouseCoopers, 2005, p. 5.
[77] See Alexander, 1998, pp. 9 ff.
[78] See Goldberg/ Sander/ Rogers, 1999, p. 3.

4.1 Alternatives to 'classic' negotiations   143

advantages and disadvantages which go along with their choice.[79] To find the right one, though, is difficult, as there are many steps between the extremes.[80] Here the instruments will be analysed in the order that the third party exercises his authority in finding a solution for the actors. Figure 5 pictures three clusters: no influence of the third party, varying influence of the third party or (full) control of the third party over the actors.[81] For all these three variants the actors (can) set up their own BATNA as a scale to be able to assess the instrument for their situation.

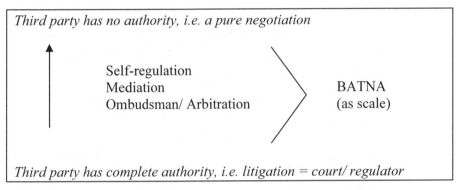

Figure 5: Loss of individual authority over the negotiated solution

Source: own design

The results which each instrument promises in negotiations where the power is distributed asymmetrically will now be discussed.

Self-regulation – Limiting external influence. Self-regulation describes a negotiation approach with self-binding mechanisms where the actors try to exclude external influence,[82] because regulation is only seen as a second best solution. The difference between the forms of self-regulation lies in the amount of authority that third parties have in the negotiation: the three different types of self-regulation are voluntary, i.e. independent of government (also known as co-regulation or alternative regulation), state required in order to establish controls, and mandatory, i.e. where certain rules are enforced by the industry but prescribed by the state.[83] For all three varia-

---

[79] See PriceWaterhouseCoopers, 2005, p. 5.
[80] See Lewicki et al., 1994, p. 352.
[81] See Raiffa, 1982, pp. 22 ff.
[82] See Stuhmcke, 2002, p. 76; Wentzel, 2002.
[83] See Latzer et al., 2003, p. 150.

tions the base element is a moderated process, where the moderation is either assumed by a member of the group or an external third party.[84]

In a self-regulatory regime the actors set up their own **framework** within which the actors can move relatively freely, but with the implicit threat that if self-regulation is not implemented and executed, institutional regulation will take its place. Self-regulation thus offers the actors the possibility of shaping their own framework, limiting the threat of regulatory uncertainty.[85] The combination of a self-regulatory regime with sunset regulation could create additional transparency.[86] In effect this allows for flexible and dynamic negotiations on different levels – within certain limits – and enables actors to **'expand the cake'**.[87] Hence there is less government failure to be expected and the flexibility of the mechanism is greater.[88] Self-regulation thus gives the (negotiation) power back to the actors and lets them focus on the content of the negotiations instead of having to worry about regulatory issues, and could therefore make self-regulation a concept superior to government regulation.[89]

In the attempt to ensure **cost** transparency and to eliminate informational asymmetries in essential facility industries the focus is traditionally placed on the role of regulation.[90] It is claimed, though, that telecommunication companies, for e.g., would have introduced many more services had governmental regulation not been in place.[91] The efficiency of decentralized systems such as self-regulation is in this respect valued more highly than that of centralized regulatory regimes: self-regulatory regimes such as ICANN give the actors a better opportunity to create valuable options and solutions in the negotiation process.[92] In fast moving markets especially

---

[84] See Poel, 2002.
[85] See Ernst & Young, 2003, claiming that regulatory change poses one of the biggest threats to companies in network markets.
[86] See Freytag/ Winkler, 2004; Deutscher Bundestag, 2004b, p. 828.
[87] See Wentzel, 2002, p. 6.
[88] See Better Regulation Task Force, 2000, p. 25.
[89] See Miller, 1985, p. 897; Ovum, 2002, gives an example of self regulation on International Roaming issues; B.A.U.M./ Difu, 2003; Difu/ VZ NRW/ WIK, 2005, are examples for reports on self regulation within the mobile telecommunications sector; Oftel, 2000, discusses the implementation of ADR mechanism
[90] See Stefanadis, 2003, p. 8.
[91] See Vogelsang, 2002, p. 5. He cites regulation of the FCC on the introduction of cellular mobile telephony, which took 10 years, as an example for this fact. The slow introduction is reported to have cost the American economy billions of dollars.
[92] See Leib, 2002.

self-regulatory approaches promise good negotiation results.[93] Even regulatory institutions find self-regulatory mechanisms, such as multilateral negotiations (see the example of the AKNN above), advantageous in terms of creativity.[94] Self-regulation could thus be the instrument which is best suited for (multilateral) negotiation situations in network markets.[95]

In a self-regulatory regime actors develop options for possible solutions together and then make their strategic choices based on common criteria, either developed together or accepted as given. That way the actors win by cooperating instead of through having to rely on a regulator.[96] There are also drawbacks, though. Some actors do not have the necessary resources to actively take part in self-regulatory regimes and are hence cut off from the information flow. The examination of the AKKN example shows that some actors are therefore prevented from profiting from self-regulatory regimes. This means that the self-regulatory regime might be a cartel.[97] Those that feel hampered by the existing (over-) regulation could thus have a strategic incentive to call for self-regulation, hoping thereby to avoid more regulation or, indeed, any regulation at all. Self-regulatory regimes might also run the danger of taking on too many free-riders, who at some stage will cause the self-regulatory system to collapse.[98] These drawbacks could be alleviated by a regulator actively taking part in such a self-regulatory regime.[99] As has already been shown in the AKNN example, self-regulation is an instrument which 'grows' with its market from 'government to governance'[100]; this is a time-phasing effect.[101]

To summarise: a self-regulatory regime is not to be mistaken for a *laissez faire* approach. Instead it needs support, especially in the beginning,[102] and active control must be exercised by a regulatory institution.[103] This is the case because self-regulatory regimes do not work in every situation, as there must be a level of willingness on the part of the actors to use the

---

[93] See Schönborn/ Wiebusch, 2002, pp. 66 ff.
[94] See Racine/ Auer, 2001.
[95] See Sicker/ Lookabaugh, 2004, pp. 26-28.
[96] See Stefanadis, 2003, p. 6.
[97] See Miller, 1985, p. 900. See also Wentzel, 2002, p. 10. On the issue of cartels see also Part IV, 2.3 below.
[98] See Schmidt, 2005b, p. 19, for the example of Skype, a company which 'acts' instead of 'working on' or 'waiting' for regulatory issues to be resolved.
[99] See Hillebrand/ Kohlstedt/ Strube-Martins, 2005.
[100] See Latzer et al., 2003, p. 150.
[101] See Marsden, 2005; Difu/ VZ NRW/ WIK, 2005, p. 8.
[102] See Sicker/ Lookabaugh, 2004, p.15.
[103] See Wentzel, 2002, p. 25; Miller, 1985, p. 903

available instruments.[104] It is therefore not that self-regulation makes regulation as such unnecessary, but rather that these two instruments co-exist and depend on each other. A combination of formal legislative instruments to supplement the role of self regulation could help to give the actors **real power**, especially in dynamic multilateral negotiation situations.[105] Comparing the instruments of self-regulation with the processes of litigation shows, though, that in network markets the mere threat of regulation alone can help the actors to find a solution even in situations where the power is distributed asymmetrically between them. This not only proves that power in negotiations depends on the source and its dynamics, but this also makes it an instrument based on the Harvard Negotiation Project.

Mediation – Focus on process support. Mediation is a negotiation approach where an independent third party supports the actors; it is still the actors, though, that make the decisions as the mediator does not have the authority to make decisions for them.[106] In other words, mediation concerns the negotiation process but not the content level.[107] Mediation is one of the oldest conflict resolution mechanisms and one which is used, to a varying extent, worldwide.[108] Thus there can be said to be not just one but a range of mediation approaches, which are applicable to different situations.[109] In some countries it is, for example, sanctioned by society that disagreements are taken care of by third parties; in others the **framework** orders disputants to make use of mediation.[110] Whilst the actual decision to make use of the instrument of mediation is made by the individuals[111] – within the possibilities of the given framework – it is the mediator who deploys the tool-

---

[104] See Lapuerta/ Tye, 1999.

[105] See Stefanadis, 2003, p. 24.

[106] See Wall/ Lynn, 1993 and Wall/ Stark/ Standifer, 2001 for an overview of the literature and the trends in mediation-research. For a definition see Wall/ Stark/ Standifer, 2001, p. 375; Alexander, 1998, pp. 34 ff.; Goldberg/ Sander/ Rogers, 1999, p. 123; Susskind/ McKearnan/ Thomas-Larmer, 1999. For a detailed description of the mediation process see Haft/ Schlieffen, 2002; Lewicki et al., 1994, pp. 357 ff. For an overview of the most recent German literature on mediation see Kals, 2003. On business mediation see Hösl, 2004; Zuberbühler, 2004; Haynes et al., 2004.

[107] See Duss-von Werdt, 2005, p. 183.

[108] See Wall/ Stark/ Standifer, 2001, p. 370; Hauser, 2002, pp. 220 ff.; Hehn, 2002, pp. 150 ff.; Duss-von Werdt, 2005, p. 37.

[109] See Duss-von Werdt, 2005, p. 38; Hauser, 2002, pp. 230 ff.; Alexander, 1998, pp. 24 ff.

[110] See Wall/ Stark/ Standifer, 2001, p. 372; Duss-von Werdt, 2005, p. 239.

[111] See Zuberbühler, 2004, p. 185, explaining how mediation 'connects' individuals and companies.

box full of mediation-instruments, all geared to supporting the negotiation process.[112] To get the mediation going, the mediator might need to temporarily assume the role of one actor in order to 'defend' him on different levels;[113] the mediator helps to find the right level by constructing a bridge between them. The mediator gives the actors the opportunity to voice their specific needs,[114] for example regarding which levels need to be addressed or excluded. The instruments of mediation thus help to set up a framework for the individual negotiation situation by re-arranging those issues within the general framework that better fit the individual negotiation situation.[115] In a certain sense this 'secure-negotiating-room' mirrors the actors' framework of rationality[116] which they are willing to set up and share with the other actors.[117] The nett effect of this is to enable the actors to **'expand the cake'** in a negotiation by minimising uncertainty.

Even though the overall goal of negotiations is to lower the transaction **cost**, there is no single ideal solution in negotiations, but rather a host of possible outcomes. Hence there is also room to create new solutions. Unfortunately, actors have a cognitive limit when it comes to their ability to grasp complex negotiations,[118] such as those about essential facilities. The instruments of mediation support the actors in structuring their negotiation process.[119] Splitting up a negotiation into many smaller issues helps to 'break down' the power, or at least the kind of power that arises from the fact that complex issues need more 'calculating' power to solve them.[120] The support of a mediator does not necessarily mean that the negotiation will be settled in the way which the actors or mediator expect, or indeed that it will be settled at all,[121] but it does limit the level of uncertainty. The mediator 'generates' a different 'language'[122] in the negotiation situation which allows the actors to talk to each other and actually listen to what the

---

[112] See Wall/ Lynn, 1993, p. 186: "We should bear in mind that mediation to a great extent is what a mediator does."; Wall/ Stark/ Standifer, 2001, pp. 375-377.
[113] See Kals/ Kärcher, 2001, p. 25.
[114] See Haynes et al., 2004, pp. 12 ff.
[115] See Hauser, 2002, pp. 267 ff.
[116] See Bazerman/ Neale, 1992, p. 171.
[117] See Haft, 2002, pp. 202 ff.
[118] See Hauser, 2002, p. 250.
[119] See Haynes et al., 2004, p. 45.
[120] See Hauser, 2002, pp. 252 ff.
[121] See Wall/ Stark/ Standifer, 2001, p. 382.
[122] See Haynes et al., 2004, pp. 278 ff.

others have to say.[123] In doing so the mediator helps the actors to create a choice of options for possible solutions.[124]

Just as the negotiation process itself must be split up into different time phases, the process of mediation is separated into different phases as well:[125] step by step the mediator helps the actors to make the different (hidden) **strategies** visible – with the consent of the other actors.[126] To do so the mediator takes control of the communication between the actors.[127] In due course the communication evolves from a discussion between actors and mediator to become more and more a negotiation between the actors. By creating transparency, mediation actively helps to limit the shifting of power in conflict situations.[128] The mediator's job is not only to help clear the obstacles that a conflict poses but to create a cooperative environment.[129] He does this by increasing the objectivity of the discussion[130] and by supporting the actors in their joint development of common criteria.[131] Thus the mediator helps the actors in the process of finding and evaluating the best alternatives available to them.[132] On the basis of their alternatives and the common criteria the actors can then make their individual decisions on how they see their relative position to each other[133] and as to how best **'cut the cake'**.[134] In this respect mediation presents itself as a cooperative instrument[135] which is very flexible even in dynamic negotiation situations.[136]

To summarise: mediation is an adequate approach when actors are trying to achieve a forward-looking, sustainable and long-term solution. This might at first be a surprising conclusion as mediation is, compared to a litigation approach, the 'weakest' ADR tool to make use of an independent third party. Even though some see asymmetrical power distribution as a

---

[123] See Duss-von Werdt, 2005, p. 210; Ayres/ Nalebuff, 1997, p. 5.
[124] See Winterstetter, 2002, p. 511.
[125] See Hauser, 2002, pp. 236 ff.
[126] See Duss-von Werdt, 2005, pp. 226 ff.
[127] See Schelling, 1958, p. 236.
[128] See Hauser, 2002, p. 323; Haynes et al., 2004, p. 75, give an example of such a power shift in a mediation
[129] See Breidenbach, 1995, pp. 111 ff.
[130] See Ayres/ Nalebuff, 1997, pp. 32 ff.
[131] See Poitras, 2005, discussing how to set incentives so that the actors increase their cooperation.
[132] See Hauser, 2002, pp. 219 ff.
[133] See Duss-von Werdt, 2005, p. 149.
[134] See Schuppert, 1990, pp. 334 ff.
[135] See Breidenbach, 1995, p. 97.
[136] See Duss-von Werdt, 2005, p. 47; Alexander, 1998, p. 3.

challenge in a mediation[137] and feel that a mediation situation can be handled more easily when the power is distributed in a symmetrical way,[138] it has been shown here that the merging of the Harvard Negotiation Project and the Power-Matrix offers the tools needed to handle such challenging situations. The only condition that needs to be met is that both actors have at least the minimum level of power needed to take part in the negotiation.[139] Even if this condition is not met the mediator could still initiate the negotiation in order to level out differences in the power distribution.[140] Still, one has to admit that mediation is not an instrument that can help in all possible situations. This demonstrates that power in negotiations depends on the source and its dynamics. In some cases mediation could even be the reason for an asymmetrical power distribution in negotiation situations: a mediator could make strategic use of power for his own good instead of 'taking care' of the actors.[141] It will thus be important to define the role of the mediator in more detail.[142] In general, though, it can be stated that mediation empowers the individual actor, as the process is guarded by the mediator but the result is controlled by the actors,[143] allowing them to **realise power** together.

Ombudsman and Arbitration – Relinquishing the power of decision.
In both the ombudsman and the arbitration approaches third parties, which are asked by the actors to support their negotiation, acquire the power to make decisions for the actors. Addressing an **arbitrator** is an option for actors who see no other opportunity of coming to a solution in their negotiation. The arbitrator 'listens' to the arguments of both sides. Then he begins by trying to help the actors develop their own solutions; if this does not work out he can then impose a solution.[144] Addressing an arbitrator is thus almost equivalent to leaving the decision with an independent third party.[145] In contrast to a court decision, though, the actors are not bound to accept the decision. Hence there is also no need to appeal against the decision of the arbitrator if there is no acceptance of the proposed solution.[146]

---

[137] See Hauser, 2002, pp. 206 ff.; Duss-von Werdt, 2005, p. 230.
[138] See Funken, 2002; Alexander, 1998, p. 215.
[139] See Breidenbach, 1995, pp. 248 ff.
[140] See Kals, 2003, p. 11.
[141] See Duss-von Werdt, 2005, pp. 226 ff.
[142] See Bazerman/ Neale, 1992, p. 142.
[143] See Kals/ Webers, 2001, p. 12.
[144] See Raiffa, 1982, p. 23.
[145] See Winkler/ Racine, 2002.
[146] See Gadlin, 2000, p. 42. He shows that the difference between the mediator and an ombudsman also lies in the ability of the intermediary to make rulings.

The concept of an **ombudsman** – originally a Swedish concept[147] – was established in order to give 'small' actors the opportunity to 'complain' about the government, the 'big' actor in a negotiation. The concept has since been extended and is now used for a wide array of issues.[148] In general an ombudsman gives 'small' actors who feel they have been treated unfairly a point of refuge. Some of the developments seem to have departed from the original concept, though,[149] making the term ombudsman rather vague.[150] Here the ombudsman concept is seen as a subgroup of arbitration, e.g. for actors in a 'big' vs. 'small' negotiation situation.

It is especially in the regulatory context of the network markets that arbitration and ombudsman approaches can be identified as complementary adjuncts to the courts and the regulators. Arbitration is generally seen as a court support mechanism which keeps the plethora of rather small cases from going to court.[151] The ombudsman concept is an internationally accepted part of innovative regulation, especially as the privatization of sectors makes changes and adaptations of the **framework** necessary.[152] It shows that the ombudsman approach addresses in the first place private customers that have problems with large companies. In the telecommunications sector, for example, these instruments have been implemented in the UK to enable private customers to appeal (Otelo) and business customers to appeal Ofcom;[153] in Germany and Austria[154] this instrument is implemented for both kinds of customers directly with the NRA. In Australia the ombudsman has become a cornerstone of the present regulatory regime of the telecommunications sector.[155] Both the arbitrator and the ombudsman support the actors by voicing their problems in negotiations. Both tools operate only on one level in the institutional hierarchy and are thus not necessarily able to help the actors gain access to different levels of the negotiation.[156] It depends to a great extent on the individuals that represent

---

[147] See Stieber, 2000.
[148] See Verbraucher-Zentrale NRW, 2004. The so-called Schlichtungsstelle Nahverkehr is to help the users of local public transport systems obtain their rights; for another example see Euro-Info-Verbraucher e.V., 2005.
[149] See Stieber, 2000, p. 52.
[150] See Stuhmcke, 2002, p. 79.
[151] See Lewicki et al., 1994, pp. 354 ff., for an overview of the concept of arbitration.
[152] See Stuhmcke, 2002, p. 79.
[153] See Otelo, 2003; Oftel, 2003.
[154] See Ruhle/ Lichtenberger/ Kittl, 2005, p. 18.
[155] See Stuhmcke, 2002.
[156] See Hauser, 2002.

the arbitrator and the ombudsman whether the actors ever get to the core of their problem and hence make room to **'expand the cake'**.

The obvious transaction **costs** caused by both an arbitrator and an ombudsman are relatively small compared to the other ADR approaches. This highlights the fact that these mechanisms were originally meant as a support device for 'small' actors that have neither the power nor the necessary resources to 'fight'. Once the actors make use of either approach they relinquish decision-making authority in return for a 'ruling'. It is not the actors that **develop options** but the neutral third party that makes use of the arguments proposed to him in order to make a decision. More so than in litigation the actors are asked to explain their cases. However, being able to voice the problem does not necessarily help to create options or solutions. The idea of the ombudsman and the arbitrator, nonetheless, is to establish a win-win situation by finding a solution which both can accept by bridging differences the actors can not resolve themselves.

The **strategy** of both the arbitration approach and the ombudsman revolves around the removal of ongoing problems so as to allow the actors to deal with more pressing issues. This includes an effort to create transparency in terms of timing and content in the problem-situation in order to be able to grasp the fundamental nature of the problem. Since the decisions cannot be predicted this leaves room for chance and this in turn could lead to dynamics. Both approaches, therefore, do not allow for a solution which necessarily fits all actors and/ or creates a cooperative environment since the goal is much more short-term oriented than the other ADR approaches.

To summarise: both the ombudsman and the arbitration approach are close to the concept of litigation. The problems that are experienced with litigation can to some extent be found here as well: in Austria, for example, 'Streitschlichtung' (i.e. arbitration) is seen as unsuccessful, since the instrument seems only to have created a lever for deliberate delays.[157] At the same time both the arbitration and ombudsman approach are to some extent the 'pre-product' for those cases where an actor does not have the prerequisite 'minimum' amount of power for negotiations.[158] In this respect it is probably true that, faced with having no chance at all of voicing discontent, arbitration approaches seem to some actors to be the embodiment of fairness.[159] If the concepts are used to bridge starting power asymmetries between the actors,[160] productive **power** can be **realised**. The problem could then be finding the right trigger for the use of these instruments, i.e.

---

[157] See Ruhle/ Lichtenberger/ Kittl, 2005, p. 18.
[158] See Stuhmcke, 2002, p. 88.
[159] See Baumol, 1986.
[160] See Stieber, 2000, p. 51.

discovering how the source and the dynamics of power correlate in such negotiation situations. The merging of the Harvard Negotiation Project and Power-Matrix, though, can be used to tackle this problem of finding the right trigger.

Can ADR mechanisms handle complex (power-)negotiations?

All of the above ADR instruments are, in one way or another, already in use in network markets and obviously all can be used in different ways, in different combinations and constellations. Most, furthermore, can be turned into 'bad' instruments if handled incorrectly, be it on purpose or by mistake. The question then is whether ADR mechanisms can handle complex (power-) negotiations and, if this is so, how. Instead of analysing a number of small cases, the following complex negotiation case – on the siting of masts for cellular networks[161] – will be 'deconstructed' to discuss the value of ADR mechanisms on the basis of the Harvard Negotiation Concept and in combination with the structure of the Power-Matrix. In this respect the following case study serves as a useful summary of the theory, the cases and the process approach developed in the present study. This case as such is not necessarily representative, but can help to provide an overview of the suitability of the results of this study. The underlying reasoning, obviously, is that ADR instruments help actors to run their negotiations efficiently.

What is the case about? The cellular telecommunications operators were granted licenses by their respective national governments to run their operations; in return they were obliged to deliver a certain mobile service coverage within a limited timeframe.[162] In order to be able to fulfil these guarantees the mobile operators must set up an extensive network-infrastructure. In contrast to the fixed network this consists of a fixed network plus highly visible masts (the radio base stations) that are sited all over the country. There are, for example, around 40,000 in the UK and around 70,000 in Germany.[163] Whereas in a fixed network every customer is connected to the network via a dedicated fixed line, in the mobile sector

---

[161] This case is based on Büllingen et al., 2004. See Büllingen/ Hillebrand/ Wörter, 2002, for more background on the discussion of issues on public exposure guidelines for radio wave frequencies. Wall/ Lynn, 1993, p. 186, complain that more data on mediation cases and tests of theories is necessary. The present study seeks to contribute to this.
[162] See Cullen International, 2005, for an international overview of these requirements. For Germany the requirement was to reach a coverage of 25% of the population by 2004 and of 50% by 2005.
[163] See Mobile Operators Association, 2004, for the build-up of a mobile network.

the last meters to connect the customer to the network are bridged with radio frequency waves.[164] The use of radio frequencies has led to major concerns about the Human Exposure to Electromagnetic Fields (EMF). In the case described here a mast was set up opposite a primary school and parents became worried about the effects of EMF on their children.

Legislation has strictly regulated, so it seems at first sight, the exact positioning and maximum emission of the masts (with the help of public exposure guidelines) in order to ensure the physical wellbeing of the public.[165] This **framework** proliferates on different levels: together with politicians on the national level 'association agreements' have been set up which specify the details of a built-up of the network and especially the masts. Even the churches and the state ministries have started to care about these issues and have designed brochures to hand out to those affected by the problem.[166] On the uppermost level laws are being set up and reviewed which try to prevent conflicts in the first place. The EU, for example, encourages the network roll-out and, besides many matters of technical detail, asks for a removal of the existing legal barriers.[167] When one looks closer, it becomes obvious that (even) this extensive regulation contains many 'grey areas' which cause uncertainty, i.e. **perceived power**. On the one hand, therefore, the framework is a cause for conflict, but at the same time a cure which helps the actors to separate and to prioritise the different issues. The actual 'battlefield' of the conflict negotiation described here is situated on the local level and centres around the question as to where exactly the mast could or should be positioned.[168] Only the approach by an independent third party[169] was sufficient to get the different actors to the table: in the case described here the representatives of the parents, the mobile operator and the city. Together with the third party, here a mediator, the key problem was elucidated, i.e. made transparent for all, and **'chains were taken off'** for further discussions; the different emphasis on the value of 'need for protection' was at the centre of this negotiation. The actors got help from the third party to start undoing the Gordian-knot by splitting their 'position' into the different underlying interests. The actors apparently had no incentive to start such a process before the offer of sup-

---

[164] See Diefenbacher et al., 2003, for a technical overview.
[165] See Scherer/ Schimanek, 2002, pp. 302 ff.
[166] See Kastenholz/ Benighaus, 2003; Diefenbacher et al., 2003. For the UK see Mobile Operators Association, 2004.
[167] See Commission of the European Communities, 2004b, p. 11.
[168] See Difu/ VZ NRW/ WIK, 2005, p. 9.
[169] The question as to whether the independent third party is really independent was actually discussed between the actors.

port from the third party. This demonstrates the fact that it is often not the content that is at the centre of negotiations and that ADR mechanisms are, obviously, able to initiate and handle such complex (power-) negotiations on the framework level.

In the public's mind the masts are equated with the exposure to electromagnetic fields and are thus a (highly) visible trigger for the problems associated with cellular telephony. The distribution of the property rights points to the fact that the mast-site is an essential facility and also makes the mobile operators the 'big' actors and the inhabitants/ parents close to the mast-site the 'small' actors. The voiced positions of both sides are rather strong but simple: mast yes or no. Because they lack access to such resources as the companies can provide, the 'small' actors are often organised in interest groups, such as local initiatives.[170] With relatively little cost the 'small' actors can hence pose a relatively serious threat to the 'big' actors.[171] In this respect such a negotiation can suddenly turn into an issue of risk management.[172] Whereas the mobile operators had initially calculated the **cost** of resistance to be part of the 'communication' budget, the opposition has grown to become a cost block of its own; this has turned out to be the obstacle for the whole roll-out process. This is **resulting power**. With the help of the external third party the actors for the first time got the opportunity to focus on the issues, instead of only having to guard against the risks they faced, for example the risk that a dynamic situation would suddenly spin out of control. Only at this stage the actors have been able to develop options, i.e. to **'create added value'**, together. The incentive for all actors to support this ADR process was thus the discovery that such options exist at all. These were, especially when compared to the available alternatives, rather promising.[173] This demonstrates that ADR mechanisms can handle complex (power-) negotiations also on the cost level.

Since the mobile operators have license obligations that they need to fulfil in time and the parents are worried about the EMF now, the timing of the activities in this conflict case has been of major importance to all actors. At the same time the amount and, even worse, the contradictory in-

---

[170] See Büllingen/ Hillebrand/ Wörter, 2002.
[171] The press activities are obviously what hurts the 'big' players the most. For an example of such an article, see Hadem, 2004, p. 38.
[172] See B.A.U.M./ Difu, 2003; Kemp/ Greulich, 2004; Difu/ VZ NRW/ WIK, 2005, p. 66.
[173] The family that owned the site where the mast was 'planted', actually received death threats.

formation concerning EMF has created uncertainty on all sides.[174] It was therefore the duty of the third party in the ADR process to keep the timing and access to content under control in order to dissolve the incentives for **strategic** moves on the part of the actors. Strategic action is governed by the different size of the actors: the negotiation-experience of a large multinational-company, here a mobile-operator, is different to that of an individual actor, for whom this kind of negotiation is beyond their daily experience. In the beginning the actors were thus not willing to cooperate at all.[175] Without the help of an external third party it is likely that the actors would have continued to 'dis-communicate'. But having already discovered possible options, created with the support of a neutral third party, the actors were able to see a solution within reach. What is left is the need to identify a solution that is acceptable to all. On the basis of objective criteria set up as a scale, which holds even when presented to outsiders;[176] overconfidence and consequent irrationality on the part of the actors is prevented.[177] In other words the actors help each other to maintain their grip on reality. The newly found options for solutions are made more 'attractive' by making the alternatives more realistic. But it is still the individual actor, in full knowledge of his own BATNA, that makes a decision to go for a certain deal or not. In the present case a contingent agreement was the basis for a package deal, i.e. **'cutting the cake'**.[178] Contingent agreements allow the actors to 'bet' on what is going to happen in the future. Postponing the resolution of issues is no longer a risk when the actors have found a way of getting along, as they usually do in the due course of an ADR process. Since the actors' expectations for the future are different, betting on them is not risky, but implies a sharing of risk. Such a step might even support the long-term 'friendship' of the actors, as such an agreement makes it necessary to negotiate again in the future:[179] the 'problem' is thus turned into a constructive agreement. This demonstrates that differences in power distribution can even be helpful in finding an innovative solution. In setting up such an agreement the actors were, obviously, ready to 'live on

---

[174] See B.A.U.M./ Difu, 2003 and Kemp/ Greulich, 2004, for an overview of the information available from the mobile operators
[175] See Büllingen/ Hillebrand/ Wörter, 2002, p. 60.
[176] In the present case the actors had to get feedback and support from the groups that they were representing and also from 'interested' actors on a constant basis.
[177] See Kahneman/ Tversky, 1991, p. 4.
[178] This paragraph builds on Bazerman/ Gillespie, 1999.
[179] See Berger et. al., 2005.

the good faith' that was built up in the process.[180] The decision for a solution that is accepted by all demonstrates **real power** and also shows that ADR mechanisms can handle complex power situations on the strategy level.

ADR delivers the tool-box for efficient power-negotiations

ADR mechanisms can support actors, even in (complex) negotiation situations where the power is distributed asymmetrically and where the source and the dynamics of power are, at least ex ante, unknown.[181] The question is whether it is proportional to claim – on the basis of this (single above) example which mainly described the application of the 'mediation' mechanism – that all ADR mechanisms are something to be striven for. To encapsulate it in one sentence: ADR mechanisms deliver the tool-box for efficient negotiations and a basis for the different kinds of third party intervention mechanisms.[182]

There are, however, also problems implementing these mechanisms. The Harvard Negotiation Project is 'only' an intuitive and easy-going concept[183] and is more of a guideline than a scientific concept.[184] There has not even been scientific research aimed at discovering to what extent it has actually led to good results.[185] This does not make it easier to convince actors to take part on a voluntary basis.[186] Hard bargainers exist, who are neither willing to stick to the guidelines set out by the Harvard Negotiation Project, nor even to consider a different approach to negotiations. On the other hand, the process guidelines are accepted and applied successfully many times every day. Success in negotiations can neither be predicted nor can it be accurately measured. This points again to the fact that the problem of **perceived** power exists even when making use of the Harvard Negotiation Project.

Obviously, many actors fear that negotiations could be unnecessarily extended by the implementation of a new process approach such as the Har-

---

[180] See Bazerman/ Gillespie, 1999, p. 7, explaining that contingent contracts are also a measure for finding out what others think. That the 'big' actor reacted after everyone else had shown that they were ready to accept the principle of such an agreement, could have meant that this was the information which the 'big' actor was waiting for.
[181] See Raiffa, 1982, p. 19.
[182] See Commission of the European Communities, 2002a.
[183] See Hättenschwiler, 2003.
[184] See Funken, 2002, p. 3.
[185] But see Hättenschwiler, 2003.
[186] See Engel, 2002b, p. 50.

vard Negotiation Project. It might at times be faster and thus even acceptable to all actors, to obtain a ruling simply in order to be able to expedite the business quickly, instead of having to negotiate. Negotiations are hence processes that consume time and resources.[187] Furthermore, the Harvard Negotiation Project is a sequential, rather than multi-tasking approach, in which one problem is handled after an other, causing results, e.g. **resulting power**, that might only be short lived.

Multilateral negotiations are common in network markets. The achievements of the Harvard Negotiation Project in negotiations with many participants, even though not much has been published (yet), are very promising:[188] the Harvard Negotiation Project can support negotiations by helping to structure the dynamics and communication between (many) actors, limiting complexity for them. This way the actors can create and then choose from common, alternative solutions. On the one hand this points to the fact that negotiation situations are much more complex than first expected, but also indicates that cooperation is the supply of **real** power.[189]

In summary, it can be stated that the Harvard Negotiation Project is not an instrument but rather a part of a support mechanism to the negotiation process, making use of the ADR toolbox. The authors of the Harvard Negotiation Project themselves note that power imbalances are seen as a problem in negotiations and hence can even be a problem for the implementation of the Harvard Negotiation Project.[190] But, as the Power-Matrix demonstrates, power depends on the perception of the other's strength and actors can counter this perception by implementing the Harvard Negotiation Project. Power definitely seems to be a challenge for such a process but nevertheless it is not a situation where these principles fail, but where they instead create real power. Together the Power-Matrix and the Harvard Negotiation Project are able to give the actors support even in difficult negotiation situations.[191]

4.2 Three steps to running efficient power-negotiations

The power distribution between the actors is one major parameter that can block the path to efficient negotiations and in that way prevent the actors

---

[187] See Hauser, 2002, p. 194.
[188] See Hillebrand/ Kohlstedt/ Strube-Martins, 2005; König/ Prader/ Zillessen, 2005, on a mediation with 54 parties on the airport in Vienna, Austria.
[189] See PriceWaterhouseCoopers, 2005, p. 5.
[190] See Fisher/ Ury/ Patton, 1991, p. 177: "Some things you can't get."; Ury, 1992.
[191] See Hauser, 2002, pp. 206 ff.

from reaching the best possible result for their individual situation. Thus a new approach to (the theory of) negotiations based on power is necessary to allow for efficient negotiations.[192] This result is not only valid for network markets, but also for other sectors in most countries.[193] Until now the active demand for such services has been limited,[194] despite the fact that there is a high supply of ADR specialists and ADR know-how.[195] To overcome this deadlock situation the potential mutual gains from trade have to be highlighted to make the use of ADR mechanisms easier.

The following three steps are proposed by which to implement, to coordinate and to initiate efficient negotiations even when power is distributed asymmetrically.

4.2.1 Step one: Implement the structural approach of the Harvard Negotiation Concept and the Power-Matrix to boost efficiency

The Harvard Negotiation Project, together with the Power-Matrix, delivers a basis on which ADR mechanisms can be set up as an efficient alternative conflict management system. There are three ways to implement instruments for handling power in (most) negotiation situations:[196] firstly, a regulatory, secondly a cost-driven, and thirdly a market-driven approach.[197]

The regulatory approach to taking off the chains with ADR. The incentives to implement the elements of the Harvard Negotiation Project call on one's selfish desire to participate in a win-win solution. Thus once ADR concepts are accepted by the actors as an efficient support for negotiations in their special situation, there is hardly any problem in keeping the actors at the table.[198] To get them there the actors need to be given the opportunity of acquiring a feeling for the toolbox of ADR mechanisms, though. Theoretically a regulator could handle this.[199] Today, so it seems, regulators and courts have insufficient incentives either to support or to switch to more efficient conflict resolution mechanisms. The ability of a regulator to refer

---

[192] See Hoffmann-Riem, 1989, pp. 15 ff.
[193] See Büllingen/ Hillebrand/ Wörter, 2002, pp. 143 ff., for an international overview of like cases.
[194] See PriceWaterhouseCoopers, 2005, p. 7; Alexander, 1998, pp. 4 ff.
[195] See Hauser, 2002, pp. 308 ff.; Flögel, 2002; Alexander, 1998; Wittschier, 2001.
[196] See Funken, 2002, pp. 6 ff.
[197] See Alexander, 1998, pp. 219 ff.
[198] See Poitras/ Bowen, 2002, p. 211; Poitras/ Bowen/ Byrne, 2003.
[199] See Engel, 2002b, p. 56.

to alternatives to litigation can ease his overload of cases.[200] Furthermore, ADR mechanisms are advantageous for regulators since there is no need for them to input substantial amounts of resources.[201] Regulators can instead position themselves as a 'helping hand' and innovator. Most of the current regulatory adaptation in network markets, though, is focused on the harmonisation of remedies, instead of designing more efficient instruments that would make remedies redundant.[202] Due to this development the need for change in dynamic network markets is seen and endorsed by almost all stakeholders.[203]

How can the problem-solving competences of ADR mechanisms be anchored within the existing (legal) framework,[204] and what needs to be done to ensure that the existing framework accepts the new instruments?[205] The regulators should make sure that the classic perspectives on how to run negotiations are not transferred on a one to one basis to the new instruments, thereby endangering their implementation right from the start.[206] Instead of such a retrograde design, transparency helps to develop a flexible, forward-looking framework for the individual situation, not limited merely to some legal or technical issues, but including precisely those issues which the actors perceive at the time to be necessary for the process. Purely legal frameworks do not give the actors an incentive to set up such a process in the first place. The contrary seems to be true: legal institutions 'invite' actors to retreat to their legal positions. Offering instead a choice of instruments improves the actors' access to their rights, supports the actors in achieving the right level in their negotiation and enhances the way the institutional system provides 'BATNAs'. Institutionalizing the availability of such a choice lowers the barriers to its implementation and thus improves the existing system as a whole. 'Checklists'[207] of criteria on how the ADR mechanisms could and should be manifested institutionally are provided by the actors.[208] Still, it is necessary that institutions support such manifestation, since not all the problems can be sufficiently tackled and solved within the existing framework.[209]

---

[200] See Poitras/ Bowen/ Byrne, 2003, pp. 257 ff.
[201] See Hauser, 2002, pp. 175 ff.
[202] See ERG, 2003b.
[203] See Beardsley/ Enriquez/ Garcia, 2004.
[204] See Holznagel, 1990; Flögel, 2002.
[205] This paragraph builds on Barendrecht, 2003.
[206] See Kals, 2003, p. 8.
[207] See ecta, 2004, pp. 14 ff.
[208] See PriceWaterhouseCoopers, 2005, pp. 15 ff.; Wentzel, 2002, p. 20.
[209] See Hoffmann-Riem, 1989, pp. VII ff.

To summarise: instead of handling all the problem issues himself the regulator can support efficient negotiations by providing room for alternative conflict management instruments for negotiations. When supplying the necessary amount of information the regulator can position ADR as an equally viable alternative to litigation. This is making use of the structural approach of the Power-Matrix and the Harvard Negotiation Project. But there is some uncertainty as to whether administrations can help to solve conflicts in negotiation situations because the institutional system, which is a source of power, does not allow for such dynamic power-handling mechanisms.[210]

The cost approach to creating added value with ADR. Every time a new process is about to be implemented – and this is true for ADR mechanisms as well – a starting investment is needed. Incomplete information complicates the actor's analysis of the predicted cost of change.[211] This establishes a barrier for dynamic change. Even though the actors know that they are best in creating new and innovative options, the necessary process cannot be implemented if the disputants have only limited access to the resources needed to do so.[212] Smaller actors might thus be more hesitant to apply new mechanisms at all; larger companies have more leeway when it comes to making use of alternative instruments.[213] At first sight the Harvard Negotiation Project therefore seems only to provide an approach for 'luxury cases'.[214]

What then makes ADR mechanisms interesting for negotiations in network markets in the cost dimension? ADR mechanisms help to reduce the complexity of negotiations by channelling information to the point where it is needed to make decisions. This eases the communication between the actors and their environment, no matter whether they be principal or agent and no matter whether power is in the game or not. The elements of the Harvard Negotiation Project allow the actors to add value to the existing choice of solutions and make even situations that initially seem to lack anything worth distributing, i.e. zero-sum games, attractive for negotiations. For example, value is already created for many actors by getting the opportunity to avoid court cases and their attendant publicity.[215] This additional value makes solving the problems of essential facilities in network markets easier.

---

[210] See Hoffmann-Riem/ Schmidt-Aßmann, 1990; Holznagel, 1990; Flögel, 2002.
[211] See Manzini/ Mariotti, 2002, p. 3.
[212] See Wall/ Stark/ Standifer, 2001, p. 383.
[213] See PriceWaterhouseCoopers, 2005.
[214] See Schuppert, 1990, p. 328.
[215] See Winterstetter, 2002, pp. 511 ff.

To summarise: the cost of changing the established negotiation process is an obstacle to implementing ADR mechanisms only when actors are not willing to invest.[216] Nevertheless, the actors can make use of the Power-Matrix and the Harvard Negotiation Project to create more value in negotiations. By projecting more value, the actors can surpass the power source and the initial hurdle of the implementation transaction cost. In addition, cost savings can be realised with ADR, in contrast to litigation, by creating more certainty with its dynamic, rather informal processes.[217]

The strategic market approach to cutting the cake with ADR. The elements of the Harvard Negotiation Project are centred around the idea that the know-how for the solution exists within the group of actors.[218] Anyone coming in from the outside to support the process of the negotiation is thus 'only' there to support or set up the network between the actors, but not to help on the content part of a solution. The problem for the implementation of such a process is that it needs to be designed – this does not mean set – before any results can be visualized.[219] A third party can bridge this part of the process by helping to clear away possible roadblocks and keeping both actors in the game until first progress is realised – by setting incentives for cooperation.[220] To be able to acquire the different inputs from the different actors in a negotiation situation a high degree of transparency is necessary. The more responsibility the actors have (or claim) for themselves, the more confidence and trust can be built up between the actors in respect to timing and content needs; this can mark the start of a long term relationship, which does not necessarily mean that the actors have to harmonise on every issue, but rather that they behave coopetitively. To rate the different options developed together the elements of the Harvard Negotiation Project help in setting up objective criteria for the individual situation. Each actor will still make his own decision between the new alternatives.

To summarise: to implement a process which initiates the cooperation between the actors it is necessary to design a transparent negotiation process. By doing this the source of power is made available for all actors to allow for a dynamic cooperation and competition on the contents of the negotiation. An ADR approach can deliver this kind of support by making use of the Power-Matrix and the Harvard Negotiation Concept. The advantages of such a coopetitive approach need to be marketed more actively, though.

---

[216] See Poitras/ Bowen, 2002, p. 224.
[217] See Breidenbach, 1995, p. 113.
[218] See Falk/ Kosfeld, 2003.
[219] See PriceWaterhouseCoopers, 2005, p. 4.
[220] See Poitras, 2005.

4.2.2 Step two: Coordinate the instruments by creating a feedback mechanism

> *"Regulatory systems call for an adequate mix of regulatory mechanisms."*
>
> Michael Latzer[221]

The problems with power in negotiations in network markets are well known and documented.[222] Why then is it so difficult to implement a new dispute resolution model which solves these problems?[223] The emergence of ADR mechanisms has obviously led to a questioning of the classic instruments and has thus created their champions and foes. As with everything that is new, these tools are seen as a threat to the existing system: the sedate sources of power, obviously, retard the new dynamics of power.[224] The established institutions worry that ADR could totally wipe away the existing established mechanisms. Others claim that because they are so complex and dynamic the new problem situations[225] require up-to-date mechanisms, even if these have not yet been thoroughly tested.[226] In both network markets some ADR instruments are already being applied;[227] the ADR Green paper of the EU commission[228] gives an overview of the many different ADR approaches in the EU. The instruments are also today being used as complements to each other, though this activity is limited to certain cases.[229]

---

[221] Latzer et al., 2003, p. 151.
[222] See Alexander, 1998, pp. 6 ff.
[223] See PriceWaterhouseCoopers, 2005, for an extensive study on this issue.
[224] See Monopolkommission, 2004b, p. 63, on the criticism of the new paragraph on mediation in the new telecommunications law. The criticism seems to be justified as the law does not go into sufficient detail to ensure that the implementation of such a new instrument can be accomplished at all; Winkler, 2005.
[225] See Schuppert, 1990, p. 330.
[226] See Dannin, 1999, p. 4. On the problems with the implementation and usage of mediation see Holznagel, 1990; Flögel, 2002; Alexander, 1998.
[227] See Holznagel, 2003, for a systematic overview of negotiation principles in the telecommunications market and the different instruments that are available, such as arbitration, mediation and the court; Deutscher Bundestag, 2004a, p. 1. "With the addition of mediation as an alternative dispute resolution instrument, unnecessary regulatory proceedings can be avoided."
[228] See Commission of the European Communities, 2002a.
[229] See Funken, 2001, who argues that even in Germany, where the courts (always) have the task of arbitrating between the actors, there is a widespread trend towards ADR procedures.

Only in the abstract realm of theory does it make sense to isolate particular instruments for particular situations. In real life it is more likely that a complex negotiation situation consists of many different events where it is best to apply a portfolio of instruments – a conflict management system – according to the exigencies of the situation and in the combinations which are found to be necessary.[230] 'Hybrid' approaches that combine the classic- with ADR instruments in negotiations are thus seen as the most promising. The challenge is then to fan out the portfolio of instruments to create a basis for efficient negotiations. To counter both the source and the dynamics of power in the negotiation there is a constant need to fine tune the instruments.[231] Together with the Power-Matrix the Harvard Negotiation Project offers both the portfolio and the mechanisms for combining the different (negotiation) instruments: a toolbox for supporting processes.

The incentives to overcome the inhibition against using 'new' mechanisms have to be established before change can become possible. The advantage of the relatively general guidelines of the Harvard Negotiation Project is that they can be used for many different markets and situations.[232] This allows the individual negotiator to adapt them to his individual negotiation situation. In markets which are strongly regulated there could be a need for a formalisation of the incentives in order to increase the implementation of the new instruments.[233] The relationship between the stakeholders in network markets evolves over time and creates learning effects, even within a negotiation.[234] Especially in a long-term relationship (or where there is a problem which is expected to recur) it could be useful to implement a learning mechanism of the kind illustrated in figure 6 below to be able to customise the toolbox for supporting processes.

---

[230] The following two paragraphs builds on PriceWaterhouseCoopers, 2005.
[231] See Ford/ Spiwak, 2003.
[232] See Leykauf, 2000.
[233] See Wall/ Stark/ Standifer, 2001, p. 385.
[234] See Coen, 2005.

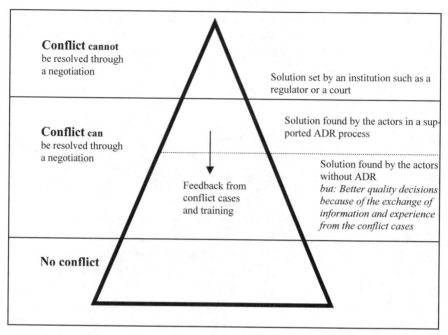

Figure 6: Learning mechanisms – customising the ADR toolbox

Source: Büllingen et al., 2004

The existing framework in network markets does not allow for a 'reset' of the whole system at once. When changes are introduced in small sequential steps, though, all actors have time to become acquainted with the new tools.[235] External help is able to further smooth such a structural transition process. In practice this could mean that a negotiation situation starts off with a moderation that can be quickly transformed into a mediated negotiation, but only if a conflict cannot be resolved by the actors themselves. If, after the (first) problem has been resolved the process were to be switched back to a regular negotiation process, the third party would revert to being a moderator instead of a mediator. Such a process respects the basic principle that the less authority is taken away from the actors, the better the actors can live with the results.

The 'feedback' process between the different levels can, over time, help to determine the appropriate level of intervention for third parties. At the same time such a feedback mechanism includes an escalation mechanism with the possibility of admitting other institutions and regulators such as the cartel office into the process. Even when a third party is not necessarily

---

[235] See Friebel/ Ivaldi/ Vibes, 2004, p. 16.

involved in the settling of a conflict or in supporting a negotiation, the option of having one on standby has an impact on the final outcome.[236] Such a 'threat effect' by creating the transparency of when and how an escalation procedure will be deployed is already part of the regulation of the network markets today: the regulatory process regards litigation only as a substitute for a failed negotiation.[237] This demonstrates that already the existence of such a feedback mechanism creates an increased level of discipline.

The regulator is under a lot of pressure, however, since it is a common complaint that the market is over-regulated. He has to tread a fine line between too much and too little regulation but at the same time has to design, or support the design of, new instruments. There is no one set of guidelines which can order the use of particular instruments in a particular situation. At first sight, the regulator's role in supporting the transfer to a 'newer' regime seems indispensable. When one looks at the network markets, though, it is difficult to understand why a new mechanism is needed to help the already "tied up and extended administrative proceedings".[238] Namely the regulator could become the new bottleneck insofar as he has no real incentive to conquer the 'old' regime. For him it is easiest to keep everything the way it is. What is needed is thus not a formal governmental proceeding but rather the opposite.[239] It is the individual actor in the negotiation who decides which instruments are to be applied. This demonstrates that the Harvard Negotiation Project mechanisms are not designed to substitute the existing tools (litigation) but to coordinate them with others – giving the actors access to the ADR toolbox. The third party is mainly necessary as a coordination mechanism. After educating the actors on how to make use of the 'new instruments', the third party can (and most of the time should) leave the actual negotiation situation to the individuals. Obviously, it is easier for the actors to handle the situation together with a third party as he can take over some of the organisational hassle of a negotiation. However, most actors still feel better making use of the ADR elements all by themselves.[240]

---

[236] See Manzini/ Mariotti, 2002, p. 19.
[237] See Wegmann, 2001, p. 256.
[238] Bruce/ Marriott, 2002, p. 2.
[239] See Bruce/ Marriott, 2002, p. 5.
[240] See Hättenschwiler, 2003, demonstrating that most of the actors consider it a good idea to freshen up their instrument-know-how after some time as the theory frequently sounds easier than it is in real life negotiation situations. One could conclude from this that actors appreciate the support they receive.

4.2.3 Step three: Initiate the use of the new tools

Steps one and two show that it is worth implementing ADR mechanisms and that it is best to set them up as a complement to the existing, classic instruments; the question remains, though, as to who initiates the (actual) change. There are three stakeholders that can initiate the change: institutions, third parties and the actors themselves.

The role of the **institutions** is to oversee (as a patron, not necessarily actively involved) the initiation and the use of the available instruments. A starting impulse from an institution is especially appreciated in situations where actors must be of a 'minimum size' in order to negotiate with the other, larger actor. This is the reason why today in conflict situations having recourse to the courts is (still) the most common, almost reflex option: if a problem in a negotiation persists most actors see the court as their first, if not only, opportunity of obtaining redress. Institutions can jump start ADR mechanisms by pointing to the fact that taking the matter to court amounts to leaving the final decision in the hands of an independent third party[241] and that there is a good chance that this decision will conflict with their interests – the classic win-lose situation. To further ease the use of ADR instruments it seems to be an obvious, if almost homeopathic, choice in fast developing (electronic) markets to make use of the electronic media for difficult negotiations.[242] A pure online mechanism creates a different kind of communication, with which every actor has first to become acquainted.[243] To be able to implement such a mechanism in an efficient manner, it is necessary that all actors are able to make use of these new instruments in the same way. Otherwise a new kind of asymmetry (a 'digital divide') would be established; it should be noted, though, that smaller actors could have an advantage in the greater speed with which they can adapt to such a new instrument. Online ADR mechanisms as such are thus not a new instrument, but variations on an old one, only making it more efficient, quicker and more flexible. In reality it will most likely complement the ADR instrument portfolio by giving easier access to the toolbox. Institutions can provide a basis for the initiation of ADR mechanisms which

---

[241] See Racine/ Winkler, 2002.
[242] This paragraph builds on Märker/ Trénel, 2003. An example of the use of online mediation is the report from Euro-Info-Verbraucher e.V., 2005. It shows that the use of online mediation instruments is suitable especially for online (network) markets.
[243] See Fietkau/ Trénel/ Prokop, 2005, p. 116.

are adapted to the individual market situation in order to ease the use of ADR.[244]

The role[245] of the **intermediary** in ADR concepts is to 'unlock' situations where the actors would not be able to achieve an efficient solution unaided.[246] It is not the job of the third party to give more power to the 'small' actor or to take away some from the 'big' actor. Third parties have a major incentive to initiate such a change because the demand for their specific know-how has been, up to now, only marginal. The danger, though, is that their only driving force could be self-interest, which would not necessarily support the actors[247] and could even result in an outcome that is no longer efficient.[248] For that reason the third parties are responsible for demonstrating their independence in the negotiation situation,[249] for example by making their work transparent.[250] There is no complete independency, though, since the mediator is, just as all the stakeholders are, a product of their environment and framework. The actors in the individual negotiation situation hence have to find out for themselves if they consider the third party to be neutral and qualified.[251] It is not necessary to regulate ADR instruments to make sure a certain quality standard is met.[252] Setting up an institute[253] to guarantee certain standards by way of education[254] could provide the impetus for the acceptance of new mechanisms, though.[255] This could be a good initiative for getting ADR off the starting blocks, and is possibly best when it comes from the third parties themselves. The role of the third party in the negotiation is to help the actors to set up a negotiation process – which includes the choice of a (qualified) third party.

When a win-win situation can be envisioned by the **actors**, it seems to be rather easy to get an ADR process started. In the best case scenario the

---

[244] See Duss-von Werdt, 2005, pp. 115 ff.
[245] See Manzini/ Mariotti, 2002, p. 2.
[246] See Wall/ Stark/ Standifer, 2001, p. 373.
[247] See Becker-Beck/ Beck/ Eberhardt, 1998, p. 109.
[248] See Manzini/ Mariotti, 2002, p. 20.
[249] See Breidenbach, 1995, p. 98; Duss-von Werdt, 2005, p. 39.
[250] See Better Regulation Task Force, 2003, pp. 8 ff.
[251] See Alexander, 1998, p. 31. She states that a mediator can never be absolutely neutral, but he works on the basis of the framework that has been set up in the mediation.
[252] See Goldberg/ Sander/ Rogers, 1999, pp. 184 ff.; Commission of the European Communities, 2002a; Commission of the European Communities, 2004c.
[253] See Alexander, 1998, p. 269.
[254] See Leykauf, 2000; Haft/ Schlieffen, 2002.
[255] See Alexander, 1998, pp. 99 ff.

actors themselves are able to define the necessary speed of their own proceedings. Such situations are easily found when the present negotiation does not have a 'long history' and offers obvious options for enlarging the 'cake' – as is often the case in negotiations on the implementation of new technologies or services. There is one major problem, however, with negotiations where several actors strike a deal together. They could behave in a collusive, i.e. illegal (cartel-style[256]) manner, so as to find solutions that benefit themselves, but are good neither for the economy as a whole nor for consumer welfare.[257] In cartel agreements a few companies 'share' the market between them. Such cases might arise from situations where actors deal with options which suit (only) the actors involved, as is the case in essential facilities markets without the involvement of a regulator.[258] This is a general problem with negotiations, though. Negotiations can always be (mis-)taken as cartel agreements between those actors striking the deal. In dynamic markets, though, and with cartel offices all over the world watching to prevent such situations, this risk is limited.[259] ADR mechanisms are thus a legal and attractive option for both individuals and companies. The approach for companies who wish to make better use of ADR mechanisms is to introduce conflict management systems before the conflict takes place, for example, by including an ADR/ mediation clause as standard in their contracts.[260] The role of those actors seeking to initiate ADR mechanisms is to market them more actively to their counterparts.[261]

4.3 Enabling support processes

Together the Power-Matrix and the Harvard Negotiation Project seem capable of giving the actors support to run negotiations efficiently even in situations where power is distributed asymmetrically – i.e. in difficult negotiation situations. Both structural mechanisms are the basis for the three steps to running efficient negotiations. These have so far covered the implementation, the coordination and the initiation of ADR by the stakeholders in negotiations. From this a basis for the promotion of efficient support processes can be designed.

---

[256] See Pies, 2000, p. 23; FAZ, 2004e, on the allegation of the EU-commission that the mobile operators arrange the terms of their roaming charges.
[257] See Tirole, 2004, p. 264.
[258] See Hoffmann-Riem, 1989, p. VII.
[259] See Korobkin/ Moffitt/ Welsh, 2004, on legal restrictions on negotiations.
[260] See PriceWaterhouseCoopers, 2005, p. 4.
[261] See Hommerich, 2005.

4.3 Enabling support processes

Figure 7 shows that realising such a concept would imply that every time an actor requests litigation, the actor would first receive mandatory information on the possibility of using ADR mechanisms as a substitute or as a complement.

```
           ↑      LITIGATION

                  ADR TOOLBOX
              No need to use it, but information is mandatory
              before one is allowed to proceed to the next step.

                  ACTORS IN NEGOTIATION SITUATION
```

Figure 7: Enabling support processes
Source: own design

The "Preliminary draft proposal for a directive on certain aspects of mediation in civil and commercial aspects" drafted by the EU-Commission describes such an automatic referral of conflicts to mediators before a court accepts a case.[262] For the network markets this could, as the example of the German telecommunications and rail transport market demonstrates, be realised with an internal instruction from the NRA to scrutinise cases firstly for their ADR suitability, before either transferring them to a third party or taking them on board. To substantiate the mandatory information, the regulator can make use of benchmarking and examples from other negotiation situations from different sectors and countries in order to demonstrate and exemplify the efficiency of the ADR mechanisms.[263] Actors, furthermore, will automatically compare the advantages of the new negotiation mechanisms with the benefits of going to court.[264] The benchmark, together with a 'little bit of story telling', will help to produce a realistic BATNA for the actors negotiating, supporting them in making the best choice of instruments in their individual situation. The major problem of the ADR instruments, their initiation,[265] would in this way be solved.

---

[262] See Commission of the European Communities, 2004a; Projektgruppe Gerichtsnahe Mediation in Niedersachsen, 2004.
[263] See Bruce/ Marriott, 2002, pp. 34 ff.
[264] See Alexander, 1998, pp. 16 ff.
[265] For that reason guidelines such as Kastenholz/ Benighaus, 2003, are only of theoretical value, as they mainly describe the mediation process as such.

This kind of design gives the regulator an incentive to actively support ADR. It will lead to a reduction of cases with the NRA and thus pave the way for faster or 'better' decisions there, without a consequent loss of authority. With such a basis for the promotion of efficient support processes it is not necessary to make the government responsible for ongoing servicing of the negotiations between the actors, but for the coordination of the process-support only.[266] One possible problem may be that such an approach will consume additional time, if the actor still decides to go to court. It is probably for this reason that theory does not yet support this idea very much.[267] Nevertheless, the method of resolution is taking advantage of the continuous, dynamic development of the framework in the network markets: a constant fine tuning of the whole support process – the tool box – to give the actors the opportunity to liberate real power in negotiations, even when power is distributed asymmetrically.

Neither the abbreviation 'ADR' nor the term 'mediation' are able to convey the full implications of these instruments: a boost of efficiency in difficult negotiation situations. Instead the term mediation, for example, is linguistically equated with rather emotional and 'soft' instruments in German-speaking countries because of its supposed relationship to the term 'meditation'.[268] The vague nature of the terms and their unfamiliar definitions[269] are thus a significant barrier to their use. Finding a new term is obviously not the only concern in implementing, coordinating and sponsoring such a new portfolio of instruments, but it is still important. I advocate the use of the term **supporting process** when speaking of the portfolio of ADR or conflict management mechanisms.[270] Such a wording points to the fact that – without explicitly naming them – the supportive instruments will be dynamically chosen and adapted in the individual negotiating situation.

In terms of practical conflict resolution and negotiation procedures there is no need to re-invent the wheel since there are well-established structural (ADR) mechanisms already available.[271] It is necessary, though, to grip the 'steering wheel' through bumpy, i.e. dynamic, terrain in network markets. To pursue the analogy further, the stakeholders should also stop discussing the new instruments and instead go for a test-drive.[272] Customers do not

---

[266] See Bruce/ Marriott, 2002, p. 45.
[267] See Wall/ Stark/ Standifer, 2001, p. 386.
[268] See Janßen, 2001, p. 37.
[269] See Hauser, 2002, p. 225; Duss-von Werdt, 2005, p. 85.
[270] See Büllingen et al., 2004; PriceWaterhouseCoopers, 2005, pp. 13 ff.
[271] See Bruce/ Marriott, 2002, pp. 77 ff.
[272] See Kohlstedt, 2003b.

appreciate the spectacle of a seemingly never-ending, academic battle on issues that have nothing to do with the product itself. In that respect customers seem to be very pragmatic – only stagnancy worries them.

What, then, are the final conclusions? Negotiations in network markets are special, but they can be handled efficiently, even if power is distributed asymmetrically between the actors. There is no obvious reason why the findings of this study should not be applied to other markets as well.

List of abbreviations

ADR	Alternative Dispute Resolution
AEG	Allgemeines Eisenbahn Gesetz (Railways Act, Germany)
AKNN	Arbeitskreis Nummerierung und Netzzusammenschaltung (Association for Numbering and Interconnection, Germany)
BATNA	Best Alternative to Negotiated Agreement
Bitkom	Bundesverband Informationswirtschaft, Telekommunikation und neue Medien e.V. (German Association for Information Technology, Telecommunications, New Media, Germany)
BNetzA	Bundesnetzagentur (formerly RegTP) (Federal Network Agency, Germany)
BRTF	Better Regulation Task force (UK)
Cable TV	Cable-Television
CEO	Chief Executive Officer
DB	Deutsche Bahn AG (formerly: Deutsche Bundesbahn and Deutsche Reichsbahn, Germany)
DSL	Digital Subscriber Line
DT	Deutsche Telekom AG (Germany)
DVWG	Deutsche Verkehrswissenschaftliche Gesellschaft e.V. (German transportation science association, Germany)
EBA	Eisenbahnbundesamt (Federal Railway Authority, Germany)
EBC	Element-Based-Costing Regime
EMF	Human Exposure to Electromagnetic Fields
EU	European Union
ICANN	Internet Corporation for Assigned Names and Numbers
IP	Internet Protocol
IPO	Initial Public Offering

MVNO	Mobile Virtual Network Operators
NRA	National Regulatory Authorities
POTS	Plain Old Telephony Service
RegTP	Regulierungsbehörde für Telekommunikation und Post (now BNetzA) (Regulatory Authority for Telecommunications and Post, Germany)
RIO	Reference Interconnect Offer
ROSCOS	Rolling Stock Companies
RR	Rail Regulator (UK)
RT	Railtrack (UK)
SMP	Significant Market Power
SMS	Short Message Service
SRA	Strategic Rail Authority (UK)
SRO	Self Regulatory Organisation
TKG	Telekommunikationsgesetz (Telecoms Acts Germany)
TOC	Train operating Companies
UK	United Kingdom
VATM	Verband der Anbieter von Telekommunikations- und Mehrwertdiensten (Operators Association of Telecoms and Value-Added Services, Germany)
VCD	Verkehrsclub Deutschland (Transport Club Germany)
VdV	Verband deutscher Verkehrsunternehmen (Association of German transport companies, Germany)
VoIP	Voice-over-IP
WLAN	Wireless Local Area Network
WLL	Wireless Local Loop
WTO	World Trade Organization
ZOPA	Zone of possible agreement

References

A.T. Kearney, 2004: Telekommunikationsunternehmen sind zu sehr mit sich selbst beschäftigt. München.

Alexander, Nadja, 1998: Wirtschaftsmediation in Deutschland. Möglichkeiten anhand internationaler, insbesondere australischer, Erfahrungen. Tübingen.

Allas, Tera/ Georgiades, Nikos, 2001: New tools for Negotiators. In: The McKinsey Quarterly, No. 2, pp. 86-97.

Allianz pro Schiene, 2004: Vorab-Stellungnahme zur Bundestagsanhörung "Zwischenbilanz und Fortführung der Bahnreform". Berlin.

Althaus, Marco, 2004: Permanent Campaigning, Consultants und Co-Kompetenzen, DIPApers 02. Deutsches Institut für Public Affairs (ed.). Potsdam.

Armstrong, Mark, 1998: Network Interconnections in Telecommunications. In: The Economic Journal, No. 108 (May), pp. 545-564.

Armstrong, Mark/ Cowan, Simon/ Vickers, John, 1994: Regulatory Reform: Economic Analysis and British Experience. Cambridge.

Arnbak, Jens, 2005: Keynote-Presentation. Presented at EURO CPR 2005, Conference March 13-15. Berlin.

Arthur D. Little, 2005: The Arthur D. Little Global Broadband Report, Update 2005. Cambridge.

Aufderheide, Detlef, 2000: Institutionelle Effizienz als Ziel der Rechtsetzung. In: Ronald Coase' Transaktionskosten-Ansatz, Pies, Ingo/ Leschke, Martin (eds.), Tübingen, pp. 141-163.

Aumann, Robert J., 1976: Agreeing to Disagree. In: Annals of Statistics, Vol. 4, No. 6, pp. 1236-1239.

Axelrod, Robert, 1988: Die Evolution der Kooperation. München.

Ayres, Ian/ Nalebuff, Barry J., 1997: Common Knowledge as a Barrier to Negotiation, Yale ICF Working Paper No. 97-01, April. Yale.

B.A.U.M./ Difu, 2003: Jahresgutachten zur Umsetzung der Zusagen der Selbstverpflichtung der Mobilfunknetzbetreiber. Berlin.

Baethge, Henning/ Hübner, Rainer, 2004: Die Fünfte Gewalt. In: Capital, No. 22, pp. 18-27.

Balsen, Werner, 2004: Bahn setzt Hoheit übers Netz als Pfand ein. In: Frankfurter Rundschau, December 28, p. 9.

Balsen, Werner, 2005: Grenzen für den Hochgeschwindigkeitszug. In: Frankfurter Rundschau, March 11, p. 10.

Balzuweit, Uwe/ Brümmer, Hinrich, 2002: Neue Aufgaben für Kommunen und Kreise: Aufgabenträger und Wettbewerb im ÖPNV. In: Nahverkehrspraxis, No. 10, pp. 16-17.
Barendrecht, Maurits, 2003: Cooperation in Transactions and Disputes: A Problem-Solving Legal System? Tilburg.
Bartos, Otomar J., 1967: How predictable are negotiations? In: The Journal of Conflict Resolution, Vol. 11, No. 4 (December), pp. 481-496.
Bauer, Antonie, 2004: Schneller als die Regulierer - Warum die Mobilfunkfirmen "freiwillig" weniger kassieren. In: Süddeutsche Zeitung, July 1, p. 19.
Baumol, William, J., 1986: Superfairness. Cambridge, Mass.
Bazerman, Max H., 1985: Norms of Distributive Justice in Interest Arbitration. In: Industrial and Labor Relations Review, Vol. 38, No. 4 (July), pp. 558-570.
Bazerman, Max H./ Gillespie, James J., 1999: Betting on the Future: The Virtues of Contingent Contracts. In: Harvard Business Review, September - October, pp. 3-8.
Bazerman, Max H./ Neale, Margaret A., 1992: Negotiating Rationally. New York.
BDI, 2004: For greater liberalisation and more private autonomy - BDI positions on the European Commission proposals for a third EU railway package. Berlin.
BDI, 2005: Privatisierung der integrierten DB AG - Auswirkungen und Alternativen. Berlin.
Beardsley, Scott/ Enriquez, Luis/ Garcia, Jon C., 2004: A New Route for Telecom Deregulation. In: The McKinsey Quarterly, No. 3, pp. 38-49.
Becker, Gary, 1983: A Theory of Competition Among Pressure Groups For Political Influence. In: The Quarterly Journal of Economics, Vol. 48, No. 3, pp. 371-400.
Becker, Gary, 1985: Public Policies, Pressure Groups, and Dead Weight Costs. In: Journal of Public Economics, No. 28, pp. 329-347.
Becker-Beck, U./ Beck, D./ Eberhardt, D., 1998: Muster und Strategien in der sozialen Interaktion zwischen Gruppen - Illustrative Interaktionsanalysen des Verlaufs zweier Erörterungstermine. In: Gruppendynamik. Theorie und Praxis. Anspruch und Wirklichkeit, Ardelt, E./ Lechner, H./ Schlögel, W. (eds.), Göttingen, pp. 96-106.
Bennemann, Stefan, 1994: Die Bahnreform - Anspruch und Wirklichkeit. Hannover.
Berg, Christoph, 1999: Beiträge zur Verhandlungstheorie. Hamburg.
Berg, Sanford V./ Foreman, R. Dean, 1996: Incentive Regulation and telco performance: a primer. In: Telecommunications Policy, Vol. 20, pp. 641-652.
Berger, Sabine et al., 2005: Mediation bei der Standortsuche von Mobilfunkanlagen. In: Multimedia und Recht, No. 7, pp. XXV-XXVII.
Berke, Jürgen, 2005: Subventionen für DSL. In: Wirtschaftswoche, No. 15, p. 10.
Berndt, A./ Kunz, M., 2000: Immer öfter ab und an? Aktuelle Entwicklungen im Bahnsektor. In: Zwischen Regulierung und Wettbewerb - Netzsektoren in Deutschland, Knieps, Günter/ Brunekreeft, Gert (eds.), Heidelberg, pp. 151-204.

Better Regulation Task Force (BRTF), 2000: Alternatives to State Regulation. London.
Better Regulation Task Force (BRTF), 2001: Economic Regulators. London.
Better Regulation Task Force (BRTF), 2004: Alternatives to Regulation. London.
Bieler, Dan, 2005: German mobile outlook 2005, Ovum. London.
Bijl, Paul de/ Peitz, Martin, 2002: Regulation and Entry into Telecommunications Markets. Cambridge.
Blount, Sally, 2002: Whoever Said that Markets Were Fair? In: Negotiation Journal, July, pp. 237-252.
Böhmer, Reinhold/ Delhaes, Daniel, 2004: Weiche Welle. In: Wirtschaftswoche, No. 50, p. 62.
Böhmer, Reinhold/ Schnitzler, Lothar, 2003: Grosses Fressen. In: Wirtschaftswoche, No. 47, pp. 46-52.
Bornshten, Keren/ Schjter, Amit M., 2003: 3G Where Art Thou? On what can and can't be learnt from the 3G Spectrum Allotment Process to-date, 1999-2003. In: Communications & Strategies, No. 50, pp. 215-235.
Bötsch, Wolfgang, 2005: Der Bahnwettbewerb gehört nicht in Stolpes Hand. In: Frankfurter Allgemeine Zeitung, March 16, p. 13.
Boulding, Kenneth E., 1963: Towards a Pure Theory of Threat Systems. In: American Economic Review, Vol. 53, No. 2 (May), pp. 424-434.
Brauner, Roman/ Kühlwetter, Hans-Jürgen, 2002: Diskriminierung ja oder nein? In: Internationales Verkehrswesen, Vol. 54, No. 10, pp. 492-495.
Breidenbach, Stephan, 1995: Mediation: Struktur, Chancen und Risiken von Vermittlung im Konflikt. Köln.
Bruce, Robert/ Marriott, Arthur, 2002: Discussion Paper on the Use of Alternative Dispute Resolution Techniques in the Telecom Sector - Draft. London.
Buchanan, James M., 1962: The Relevance of Pareto Optimality. In: The Journal of Conflict Resolution, Vol. 6, No. 4 (December), pp. 341-354.
Buchanan, James M., 1966: An Individualistic Theory of Political Process. In: Varieties of Political Theory, Easton, David (ed.), Englewood Cliffs, N.J., pp. 25-37.
Büchner, Lutz Michael, 1990: Die Neugliederung der Deutschen Bundespost - Strukturen und Konsequenzen. In: Juristische Arbeitsblätter (JA), No. 6 and 7, pp. 194-199 and 228-234.
Büchner, Lutz Michael, 1999: Post und Telekommunikation: Eine Bilanz nach zehn Jahren Reform. Heidelberg.
Büllingen, Franz et al., 2004: Alternative Streitbeilegung in der aktuellen EMVU-Debatte. Bad Honnef.
Büllingen, Franz/ Hillebrand, Annette/ Wörter, Martin, 2002: Elektromagnetische Verträglichkeit zur Umwelt (EMVU) in der öffentlichen Diskussion. Bad Honnef.
Bundeskartellamt, 2002: Tätigkeitsbereicht 2001/ 2002. BT-Drucksache 15/1226. Bonn.
Bundesministerium für Verkehr, 2001: Bericht der Task Force "Zukunft der Schiene". Bonn.

Bundesministerium für Wirtschaft und Arbeit, 2003: Benchmark - Internationale Telekommunikationsmärkte. Berlin.
Bundesministerium für Wirtschaft und Arbeit, 2005: www.zukunft-breitband.de, press-release, July 20. Berlin.
Bundesrat, 2004: Anrufung des Vermittlungsausschusses durch den Bundesrat - Telekommunikationsgesetz (TKG). BT-Drucksache 200/04 (Beschluss). Berlin.
Bundesrat, 2005: Entwurf eines Ersten Gesetzes zur Änderung des Allgemeinen Eisenbahngesetzes. BT-Drucksache 91/05 (Beschluss). Berlin.
Bundesregierung, 2002a: Drittes Gesetz zur Änderung eisenbahnrechtlicher Vorschriften, Entwurf. Berlin.
Bundesregierung, 2002b: Koalitionsvertrag - Erneuerung, Gerechtigkeit, Nachhaltigkeit; Für ein wirtschaftlich starkes, soziales und ökologisches Deutschland. Für eine lebendige Demokratie. Berlin.
Bundesregierung, 2002c: Stellungnahme der Bundesregierung zum Tätigkeitsbericht der Regulierungsbehörde für Telekommunikation und Post 2000/2001 und zum Sondergutachten der Monopolkommission. Berlin.
Bundesregierung, 2004: Informationsgesellschaft Deutschland 2006. Berlin.
Canoy, Marcel/ Bijl, Paul de/ Kemp, Ron, 2004: Access to telecommunications networks. In: The economics of antitrust and regulation in telecommunications: perspectives for the new European Regulatory Framework, Buigues, Pierre A./ Rey, Patrick (eds.), Cheltenham, pp. 135-168.
Carr, Albert Z., 1968: Is business bluffing ethical? In: Harvard Business Review, January, pp. 143-153.
Carraro, Carlo/ Marchiori, Carmen/ Sgobbi, Alessandra, 2005: Advances in Negotiation Theory: Bargaining, Coalitions and Fairness. Milan.
Cassel, Susanne, 2001: Politikberatung und Politikerberatung: Eine institutionenökonomische Analyse der wissenschaftlichen Beratung der Wirtschaftspolitik, Beiträge zur Wirtschaftspolitik, No. 76. Bern.
Cave, Martin, 2004: Economic aspects of the new regulatory regime for electronic communications services. In: The economics of antitrust and regulation in telecommunications: perspectives for the new European Regulatory Framework, Buigues, Pierre A./ Rey, Patrick (eds.), Cheltenham, pp. 27-41.
Charles, Mirjaliisa, 1995: Organisational power in business negotiations. In: The Discourse of Business Negotiation, Ehlich, Konrad/ Wagner, Johannes (eds.), Berlin, pp. 151-174.
Coase, Ronald H., 1937: The Nature of The Firm. In: Economica, Vol. 4, pp. 386-405.
Coase, Ronald H., 1960: The Problem Of Social Cost. In: The Firm, The Market, and The Law, Chicago, pp. 95-156.
Coase, Ronald H., 1984: The New Institutional Economics. In: Journal of Institutional and Theoretical Economics, Zeitschrift für die gesamte Staatswissenschaft (ZgS), No. 140, pp. 229-231.
Coase, Ronald H., 1988: The Firm, The Market and The Law. Chicago.
Coddington, Alan, 1966: A Theory of the Bargaining Process: Comment. In: American Economic Review, Vol. 56, No. 3 (June), pp. 522-530.

Coddington, Alan, 1968: Theories of the Bargaining Process. London.
Coen, David, 2005: Managing the Political Life Cycle of Regulation in the UK and German Telecommunication Sectors. In: Annals of Public and Cooperative Economics, Vol. 76, No. 1 (March), pp. 59-84.
Commission of the European Communities, 2002a: Green Paper on alternative dispute resolution in civil and commercial law, COM (2002) 196 final. Brussels.
Commission of the European Communities 2002b: Directive 2002/19/EC of the European Parliament and of the Council of 7 March 2002 on access to, and interconnection of, electronic communications networks and associated facilities (Access Directive). Brussels.
Commission of the European Communities, 2003: Green paper on services of general interest, COM (2003) 270 final. Brussels.
Commission of the European Communities, 2004a: Draft European Code of Conduct for Mediators. Brussels.
Commission of the European Communities, 2004b: Mobile Broadband Services, COM (2004) 447. Brussels.
Commission of the European Communities, 2004c: Preliminary draft proposal for a directive on certain aspects of mediation in civil and commercial aspects. Brussels.
Commission of the European Communities, 2004d: European Electronic Communications Regulation and Markets 2004, COM (2004) 759 final. Brussels.
Commission of the European Communities, 2005a: Case DE/2005/0150: Wholesale unbundled access (including shared access) to metallic loops and subloops for the purpose of providing broadband and voice services in Germany. Brussels.
Commission of the European Communities, 2005b: Commission Recommendation of 06/IV/2005 on broadband electronic communications through powerlines, C (2005) 1031 final. Brussels.
Conrath, David W., 1970: Experience as a Factor in Experimental Gaming Behavior. In: The Journal of Conflict Resolution, Vol. 14, No. 2 (June), pp. 195-202.
Crandall, Robert W./ Sidak, J. Gregory, 2004: Should Regulators Set Rates to Terminate Calls on Mobile Networks. In: Yale Journal on Regulation, Vol. 21, pp. 1-46.
Crandall, Robert W./ Sidak, J. Gregory/ Singer, Hal J., 2002: The Empirical Case Against Asymmetric Regulation of Broadband Internet Access. In: Berkeley Technology Law Journal, Vol. 17, No. 1, pp. 953-985.
Cross, John G., 1966: A Theory of the Bargaining Process: Reply. In: American Economic Review, Vol. 56, No. 3 (June), pp. 530-533.
Cross, John G., 1969: The Economics of Bargaining. New York.
Cullen International, 2005: Cross-Country-Analysis, June. Namur.
Dahl, Robert A., 1957: The Concept of Power. In: Behavioral Science, Vol. 2, No. 3 (July), pp. 201-215.
Dannin, Ellen J., 1999: Contracting Mediation: The Impact of Different Statuatory Regimes. San Diego.

Demsetz, Harold, 1988: Why Regulate Utilities? In: Chicago Studies in Political Economy, Stigler, George J. (ed.), Chicago, pp. 267-278.

Deregulierungskommission, 1991: Marktöffnung und Wettbewerb. Stuttgart.

Deutsch, Morton / Krauss, Robert M., 1962: Studies of Interpersonal Bargaining. In: The Journal of Conflict Resolution, Vol. 6, No. 1 (March), pp. 52-76.

Deutsche Bank, 2004: The Hotline: Where's the fixed-to-mobile migration? Frankfurt.

Deutsche Telekom, 2004: Das neue TKG - Position der Deutschen Telekom AG. Bonn.

Deutscher Bundestag, 2003: Behandlung von virtuellen Netzen in der Novelle des Telekommunikationsgesetzes. BT-Drucksache 15/1449. Berlin.

Deutscher Bundestag, 2004a: Erste Beratung des von der Bundesregierung eingebrachten Entwurfs eines Telekommunikationsgesetzes (TKG). Plenarprotokoll, 86. Sitzung. Berlin.

Deutscher Bundestag, 2004b: Ausschuss für Wirtschaft und Arbeit, Wortprotokoll 49. Sitzung. Protokoll 14/49. Berlin.

Dialog Consult/ VATM, 2004: Sechste gemeinsame Marktanalyse zur Telekommunikation. Köln.

Die Bahn, 2002a: Expansionsverhalten der Telekom bei Marktöffnung. Berlin.

Die Bahn, 2002b: Bericht des Wettbewerbsbeauftragten. Berlin.

Die Bahn, 2003: Bericht des Wettbewerbsbeauftragten. Berlin.

Die Bahn, 2004: Bericht des Wettbewerbsbeauftragten. Berlin.

Die Bahn, 2005a: Die Bahn auf Kurs. Berlin.

Die Bahn, 2005b: Wettbewerbsbericht 2005. Berlin.

Diefenbacher, Hans et al., 2003: Mobilfunk auf dem Kirchturm. Informationen und Entscheidungshilfen für Kirchengemeinden. Iserlohn.

Difu/ VZ NRW/ WIK, 2005: Jahresgutachten zur Umsetzung der Zusagen der Selbstverpflichtung der Mobilfunknetzbetreiber. Berlin.

Donegan, Michelle, 2005: Free at last - VoIP Peering. In: Total Telecom Magazine, May, p. 29.

Donges, Juergen B./ Freytag, Andreas, 2004: Allgemeine Wirtschaftspolitik. Stuttgart.

Drahos, Peter, 2003: When the weak bargain with the strong: Negotiations in the WTO. Canberra.

Druckman, Daniel, 1971: The Influence of the Situation in Interparty Conflict. In: The Journal of Conflict Resolution, Vol. 15, No. 4 (December), pp. 523-554.

Duss-von Werdt, Joseph, 2005: homo mediator. Stuttgart.

EBA, 2004: Kompetenz und Verantwortung - 10 Jahre Eisenbahn-Bundesamt. Bonn.

Eberlein, Burkhard, 2001: To Regulate or not to Regulate Electricity: Explaining the German Sonderweg in the EU Context. In: Journal of Network Industries, No. 22, pp. 353-384.

Eberlein, Burkhard/ Grande, Edgar, 2000: Regulation and Infrastructure Management: German Regulatory Regimes and the EU Framework. In: German Policy Studies / Politikfeldanalyse, Vol. 1, No. 1, pp. 39-66.

ecta, 2004: Regulatory Scorecard - Report on the relative effectiveness of the regulatory frameworks for electronic communications in Belgium, Denmark, France, Germany, Ireland, Italy, the Netherlands, Spain, Sweden and the United Kingdom. London.
Egger, Ueli, 2002: Training Harvard Konzept. Zürich.
Ehlich, Konrad/ Wagner, Johannes, 1995: The Discourse of Business Negotiation. Ehlich, Konrad/ Wagner, Johannes (eds.), Berlin, pp. 1-8.
EITO, 2005: European Information Technology Observatory. Berlin.
Elixmann, Dieter/ Metzler, Anette/ Schäfer, Ralf G., 2004: Kapitalmarktinduzierte Veränderungen von Unternehmensstrategien und Marktstrukturen im TK-Markt, WIK-Diskussionsbeiträge, No. 251. Bad Honnef.
Elixmann, Dieter/ Schäfer, Ralf G./ Schwab, Rolf, 2004: Der Markt für Auskunfts- und Mehrwertdienste in Deutschland - Wirtschaftliche Bedeutung und Kundennutzen. Bad Honnef.
Emerson, Richard M., 1962: Power-Dependence Relations. In: American Sociological Review, Vol. 27, No. 1 (February), pp. 31-41.
Engel, Christoph, 2002a: European Telecommunications Law: Unaffected by Globalisation?, Preprints aus der Max-Planck-Projektgruppe Recht der Gemeinschaftsgüter, No. 3. Bonn.
Engel, Christoph, 2002b: Verhandelter Netzzugang, Preprints aus der Max-Planck-Projektgruppe Recht der Gemeinschaftsgüter, No. 4. Bonn.
EON, 2005: Anreizregulierung zur nachhaltigen Produktivitätssteigerung der deutschen Strom- und Gasnetze: Das Produktivitätsteigerungsmodell ("Pro+" Modell) von EON. Düsseldorf.
Epstein, Gil S./ Nitzan, Shmuel, 2005: Lobbying and Compromise. In: CESIfo Working Paper No. 1413, February, München.
ERG, 2003a: ERG-IRG Work Programme 2004. Brussels.
ERG, 2003b: Consultation Document on a Draft Joint ERG/EC approach on appropriate remedies in the new regulatory framework as of 21/11/2003. Brussels.
Erlei, Mathias, 1998: Institutionen, Märkte und Marktphasen: allgemeine Transaktionskostentheorie unter spezieller Berücksichtigung der Entwicklungsphasen von Märkten. Tübingen.
Erlei, Mathias/ Leschke, Martin/ Sauerland, Dirk, 1999: Neue Institutionenökonomik. Stuttgart.
Ermert, Monika, 2004: Eine eigene Nummerngasse für Voice-over-IP. In: heise online, July 15.
Ernst & Young, 2003: Telecommunications? München.
Ernst, Matthias, 2001: Verlagerung der Handlungsfelder sektorspezifischer Regulierung in der Telekommunikation - Marktabgrenzung und Marktbeherrschung auf den Märkten für Sprachtelefonie in Deutschland 2001. In: Zeitschrift für Verkehrswissenschaft, Vol. 72, No. 3, pp. 176-202.
Essig, Wolfgang, 2004: Bitstream Access als notwendiges Element der Zugangsregulierung, TIM Spring Meeting, April 23. Frankfurt.
Esteban, Joan/ Sakovics, Jozsef, 2003: Endogenous bargaining power. Barcelona.

Euro-Info-Verbraucher e.V., 2005: Jahresbericht 2004 eCommerce-Verbindungsstelle Deutschland. Kehl.
Evans, John/ Fingleton, John, 2002: Entry Regulation and the Influence of an Incumbent Special Interest Group. In: CESIfo Working Paper No. 787. München.
Ewers, Hans-Jürgen/ Ilgmann, Gottfried, 2001a: Trassenpreissystem TPS 01: Kurzgutachten im Auftrag der Connex und der Hessischen Landesbahn GmbH. Berlin and Hamburg.
Ewers, Hans-Jürgen/ Ilgmann, Gottfried, 2001b: Zukunft des Schienenverkehrs - Gutachten im Auftrag der FDP-Bundestagsfraktion. Berlin and Hamburg.
Falk, Armin/ Kosfeld, Michael, 2003: It's all about Connections: Evidence on Network Formation, Institute for Empirical Research in Economics, Working Paper Series, No. 146. Zürich.
Fant, Lars, 1995: Negotiation discourse and interaction in a cross-cultural perspective: The case of Sweden and Spain. In: The Discourse of Business Negotiation, Ehlich, Konrad / Wagner, Johannes (eds.), Berlin, pp. 177-201.
FAZ (Frankfurter Allgemeine Zeitung), 2002a: DB verknüpft Netzausbau mit Verträgen. In: Frankfurter Allgemeine Zeitung, April 16, p. 15.
FAZ (Frankfurter Allgemeine Zeitung), 2002b: Konkurrenz zum neuen ICE Frankfurt-Köln scheitert. In: Frankfurter Allgemeine Zeitung, June 28, p. 67.
FAZ (Frankfurter Allgemeine Zeitung), 2003a: Deutsche Bahn verliert langsam Marktanteil. In: Frankfurter Allgemeine Zeitung, February 14, p. 11.
FAZ (Frankfurter Allgemeine Zeitung), 2003b: Die Bahn öffnet ihr Stromnetz und erhöht die Preise. In: Frankfurter Allgemeine Zeitung, November 15.
FAZ (Frankfurter Allgemeine Zeitung), 2004a: Unter Verdacht - Bahnchef Mehdorn und Berater Meyer unter Korruptionsverdacht. In: Frankfurter Allgemeine Zeitung, March 17, pp. 13 and 15.
FAZ (Frankfurter Allgemeine Zeitung), 2004b: Bund und EU wollen mehr Bahnwettbewerb. In: Frankfurter Allgemeine Zeitung, March 18, p. 13.
FAZ (Frankfurter Allgemeine Zeitung), 2004c: Am Börsengang der Bahn scheiden sich die Geister. In: Frankfurter Allgemeine Zeitung, March 30, p. 13.
FAZ (Frankfurter Allgemeine Zeitung), 2004d: Kabel Deutschland gerät beim Kartellamt unter Druck. In: Frankfurter Allgemeine Zeitung, July 26, p. 14.
FAZ (Frankfurter Allgemeine Zeitung), 2004e: EU-Kartellverfahren gegen Vodafone und MMO2 - Verdacht der Preisabsprache bei Roaming-Gebühren. In: Frankfurter Allgemeine Zeitung, July 27.
FAZ (Frankfurter Allgemeine Zeitung), 2005a: Regulierer soll Schienennetz überwachen. In: Frankfurter Allgemeine Zeitung, January 24, p. 11.
FAZ (Frankfurter Allgemeine Zeitung), 2005b: Connex heizt den Preiswettbewerb auf der Schiene an. In: Frankfurter Allgemeine Zeitung, January 26, p. 11.
FAZ (Frankfurter Allgemeine Zeitung), 2005c: Die Bahn bekommt eine eigene Regulierungsbehörde. In: Frankfurter Allgemeine Zeitung, February 2, p. 13.
FAZ (Frankfurter Allgemeine Zeitung), 2005d: Bahn-Konkurrenten reicht neue Wettbewerbsaufsicht nicht. In: Frankfurter Allgemeine Zeitung, March 16, p. 13.

FAZ (Frankfurter Allgemeine Zeitung), 2005e: Deutsche Bahn verzeichnet 250 Millionen Euro Gewinn. In: Frankfurter Allgemeine Zeitung, March 17, p. 19.

FAZ (Frankfurter Allgemeine Zeitung), 2005f: In Bonn entsteht ein Super-Regulierer. In: Frankfurter Allgemeine Zeitung, March 18, p. 14.

FAZ (Frankfurter Allgemeine Zeitung), 2005g: Vivendi fordert 2,2 Milliarden Euro von der Telekom. In: Frankfurter Allgemeine Zeitung, May 18, p. 18.

FAZ (Frankfurter Allgemeine Zeitung), 2005h: Die 100 größten Unternehmen. In: Frankfurter Allgemeine Zeitung, July 5.

FDP, 2004: Fakten zur Bahnreform. Berlin.

Fietkau, Hans-Joachim/ Trénel, Matthias/ Prokop, Juliane, 2005: Kommunikationsmuster in (online) mediierten Diskursen. In: Zeitschrift für Konfliktmanagement, No. 4, pp. 115-118.

Fisher, Roger/ Ury, William L./ Patton, Bruce, 1991: Getting to yes: negotiating agreement without giving in. Harvard.

Flögel, Thomas D., 2002: Die Mediation im nationalen und internationalen Wirtschaftsverkehr. Köln.

Fockenbrock, Dieter, 2005: Die Maschinisten vom "Einstein". In: Handelsblatt, January 28, p. 24.

Ford, George S./ Spiwak, Lawrence J., 2003: Set it and forget it? Market Power and the consequences of premature deregulation in telecommunications markets, Phoenix Center Policy Paper No. 18 (July). Phoenix.

Foster, Pacey C., 2002: Negotiations in embedded relationships: Toward an integrative framework. Boston.

Foucault, Michel, 1977: Discipline and Punish: The Birth of the Prison. London.

Franz, Ulrich, 1968: Theorie der Verhandlung: Versuch einer Analyse des Verhandlungsprozesses in bilateralen und multilateralen Strukturen unter Einschluß der soziologischen, sozial-psychologischen und politischen Elemente des Konflikts. Freiburg.

Fredebeul-Krein, Markus/ Freytag, Andreas, 1999: The case for a more binding WTO agreement on regulatory principles in telecommunication markets. In: Telecommunications Policy, Vol. 23, pp. 625-644.

Frey, Bruno S., 1991: The Public Choice View of International Political Economy. In: The Political Economy of International Organizations: A Public Choice Approach, Vaubel, Roland/ Willett, Thomas D. (eds.), Boulder, Colorado, pp. 7-26.

Freytag, Andreas, 1998: Fluch oder Segen von Interessengruppen? In: Gary Beckers ökonomischer Imperialismus, Pies, Ingo/ Leschke, Martin (eds.), Tübingen, pp. 232-237.

Freytag, Andreas/ Winkler, Klaus, 2003: Entgeltregulierung in der Telekommunikation: Überlegungen für ein neues Konzept. In: Wirtschaftsdienst, No.10, pp. 677-684.

Freytag, Andreas/ Winkler, Klaus, 2004: The Economics of Self-Regulation in Telecommunications under Sunset Legislation, Jenaer Schriften zur Wirtschaftswissenschaft, No. 17. Jena.

Friebel, Guido/ Ivaldi, Marc/ Vibes, Catherine, 2004: Railway (De)Regulation: A European Efficiency Comparison, CEPR Discussion Paper No. 4319. London.

Frieden, Rob, 2004: Regulatory Arbitrage Strategies and Tactics in Telecommunications. Pennsylvania.
Friedrich, Horst, 2003: So wird die Bahn kein normales Unternehmen. In: Frankfurter Allgemeine Zeitung, September 27, p. 12.
Frost&Sullivan, 2004: Frost&Sullivan's Analysis Of The European Fixed-Mobile Substitution (Report B283). New York.
Frühbrodt, Lutz, 2002a: Telekommunikation: "Die Schnellen fressen die Langsamen". In: Die Welt, September 9, p. 15.
Frühbrodt, Lutz, 2002b: Interview Harald Stöber: "Der Regulierer sollte endlich ein klares Zeichen setzen". In: Die Welt, November 25, p. 14.
Funken, Katja, 2001: Comparative Dispute Management: Court Connected Mediation in Japan and Germany. München.
Funken, Katja, 2002: The Pros and Cons of Getting to Yes. München.
Furubotn, Eirik G./ Richter, Rudolf, 1997: Institutions and Economic Theory - The Contributions of the New Institutional Economics. Ann Arbor.
Gadlin, Howard, 2000: The Ombudsman: What's in a Name? In: Negotiation Journal, January, pp. 37-48.
Germis, Carsten, 2004: Die Bahn-Netzwerker. In: Frankfurter Allgemeine Zeitung, March 21, p. 45.
Ghertman, Michel/ Quélin, Bertrand, 1995: Regulation and transaction costs in telecommunications: A research agenda. In: Telecommunications Policy, Vol. 19, pp. 487-500.
Gibbons, Robert, 1992: A Primer in Game Theory. New York.
Gifford, Raymond L., 2003: Regulatory Impressionism: What Regulators Can and Cannot Do. In: Review of Network Economics, Vol. 2, No. 4, pp. 466-479.
Giovannetti, Emanuele, 2005: Diagonal Mergers and Foreclosure in the Internet. In: Review of Network Economics, Vol. 4, No. 1, pp. 33-62.
Glachant, Jean-Michel, 2002: Why Regulate Deregulated Network Industries? In: Journal of Network Industries, No. 3, pp. 297-311.
Goldberg, Stephen B./ Sander, Frank E. A./ Rogers, Nancy H., 1999: Dispute Resolution: negotiation, mediation, and other processes. New York.
Golvin, Charles S., 2004: Cord-Cutting Goes Mainstream, Forrester Research. Cambridge.
Grabitz, Ileana, 2003: Connex beklagt Imageverlust der Bahn. In: Financial Times Deutschland, June 26.
Grass, Günther, 2005: Freiheit nach Börsenmaß. In: Die Zeit, May 4, pp. 1-2.
Greene, Robert, 2000: The 48 Laws of Power. London.
Hadem, Björn, 2004: "Studie verharmlost Strahlungsgefahr". In: Frankfurter Rundschau, July 8, p. 38.
Haft, Fritjof, 1992: Verhandlung. München.
Haft, Fritjof/ Schlieffen, Katharina Gräfin von, 2002: Handbuch Mediation. München.
Haley, Jay, 2002: Die Jesus-Strategie. Heidelberg.
Handelsblatt, 2001: Kurth: UMTS-Infrastruktursharing keine Lizenzregeländerung. In: Handelsblatt, June 20.
Handelsblatt, 2005a: Preise senken. In: Handelsblatt, February 11, p. 11.

Handelsblatt, 2005b: Streit um Regulierer für das Schienennetz. In: Handelsblatt, February 16, p. 4.
Handelsblatt, 2005c: Experten lehnen seperate Bahnaufsicht ab. In: Handelsblatt, February 24, p. 4.
Harsayni, John C., 1956: Approaches to the Bargaining Problem Before and After the Theory of Games: A Critical Discussion of Zeuthen's, Hicks' and Nash's Theories. In: Econometrica, Vol. 24, No. 2 (April), pp. 144-157.
Harsayni, John C., 1962: Measurement of Social Power, Opportunity Costs, and the Theory of Two Person Bargaining Games. In: Behavioral Science, Vol. 7, No. 1 (January), pp. 67-80.
Hart, Jeffrey, 1976: Three Approaches to the Measurement of Power in International Relations. In: International Organization, Vol. 30, No. 2 (Spring), pp. 289-305.
Hättenschwiler, Christian, 2003: Das Harvard-Konzept: Erfahrungen mit erfolgreichen Verhandlungen - Empirische Forschungsberichte zum Einsatz des Konzepts. Diplomarbeit, Zürcher Hochschule Winterthur. Zürich.
Hauser, Christoph, 2002: Eine ökonomische Theorie der Mediation. Luzern.
Haynes, John M. et al., 2004: Mediation - Vom Konflikt zur Lösung. Stuttgart.
Hehn, Marcus, 2002: Entwicklung und Stand der Mediation - ein historischer Überblick. In: Handbuch Mediation, Haft, Fritjof/ Schlieffen, Katharina Gräfin von (eds.), München, pp. 150-171.
Heinacher, Peter, 2004: Fördert das neue TKG den Wettbewerb in der Telekommunikation in Deutschland? In: MedienWirtschaft, No. 2.
Heinen, Anne, 2001: Access to Electricity Networks: The Application of the 'Essential Facilities Doctrine' by the German Federal Cartel Office. In: Journal of Network Industries, No. 2, pp. 385-426.
Heinrichs, Horst-Peter, 2004: Interview, January 3 in Bonn. Eisenbahn-Bundesamt (EBA), Abteilung 1.
Hempell, Thomas, 2005: Benchmark "Internationale Telekommunikationsmärkte", Zentrum für Europäische Wirtschaftsforschung (ZEW), Mannheim, im Auftrag des Bundesministeriums für Wirtschaft und Arbeit. Bundesministerium für Wirtschaft und Arbeit (ed.). Mannheim.
Heng, Stefan, 2003: Mehr als "inszenierter Wettbewerb" in der Telekommunikation, Deutsche Bank Research, Economics No. 37. Frankfurt.
Heng, Stefan, 2004a: Mobile telephony - cooperation and value-added are key to further success, Deutsche Bank Research, Economics No. 42. Frankfurt.
Heng, Stefan, 2004b: Wegweisende Innovationen der Informations- und Kommunikationstechnologien, Deutsche Bank Research, Economics No. 46. Frankfurt.
Heng, Stefan, 2005: Breitband: Europa braucht mehr als DSL, Deutsche Bank Research, Economics No. 54. Frankfurt.
Héritier, Adrienne, 2001a: The Politics of Public Services in European Regulation, Preprints aus der Max-Planck-Projektgruppe Recht der Gemeinschaftsgüter, No. 1. Bonn.

Héritier, Adrienne, 2001b: Regulator-Regulatee Interaction in the Liberalized Utilities: Access and Contract Compliance in the Rail Sector, Preprints aus der Max-Planck-Projektgruppe Recht der Gemeinschaftsgüter, No. 12. Bonn.
Hess, Gregory D., 2003: The Economic Welfare Cost of Conflict, CESIfo Working Paper No. 852. München.
Heun, Sven-Erik et al., 2002: Handbuch Telekommunikationsrecht. Köln.
Hillebrand, Annette/ Kohlstedt, Alexander/ Strube-Martins, Sonja, 2005: Selbstregulierung bei Standardisierungsprozessen am Beispiel von Mobile Number Portability, WIK-Diskussionsbeitrag No. 266. Bad Honnef.
Hirschfeld, Oliver, 2003: Interview, February 5, 2003 in Frankfurt. Deutsche Bahn AG, Recht Personenverkehr.
Hirschhausen, Christian von/ Siegmann, Jürgen, 2004: Synetra: Synergien zwischen Bahnnetz und -transport: Praxis, Probleme, Potentiale, Bericht für das BMBF. Berlin.
Höckels, Astrid/ Rottenbiller, Silvia, 2003: Entbündelung, Betreiberauswahl im Ortsnetz, Resale - ein sinnvolles Potpurri an Regulierungsinstrumenten zur Förderung von Wettbewerb im Ortsnetz? In: Wirtschaft und Wettbewerb, No 12, pp. 1290-1294.
Hoffmann-Riem, Wolfgang, 1989: Konfliktmittler in Verwaltungsverhandlungen, Forum Rechtswissenschaft, Beiträge zu neueren Entwicklungen in der Rechtswissenschaft, No. 22. Heidelberg.
Hoffmann-Riem, Wolfgang/ Schmidt-Aßmann, Eberhard (Hrsg.), 1990: Konfliktbewältigung durch Verhandlungen Vol. 1, Informelle und mittlerunterstützte Verhandlungen in Verwaltungsverfahren. Baden-Baden.
Holznagel, Bernd, 1990: Konfliktlösung durch Verhandlungen: Aushandlungsprozesse als Mittel der Konfliktverarbeitung bei der Ansiedlung von Entsorgungsanlagen für besonders überwachungsbedürftige Abfälle in den Vereinigten Staaten und der Bundesrepublik. Baden-Baden.
Holznagel, Bernd, 2003: Studie zu den Anforderungen an die gerichtliche Kontrolle im Lichte der TKG-Novellierung. Münster.
Hommerich, Christoph, 2005: Marketing als Vertrauensbildung – Herausforderungen an das Marketing der Mediatoren. In: Zeitschrift für Konfliktmanagement, No. 3, pp. 94-98.
Hösl, Gerhard G., 2004: Mediation - die erfolgreiche Konfliktlösung: Grundlagen und praktische Anwendung. München.
Huntington, Samuel P., 1996: Kampf der Kulturen: Die Neugestaltung der Weltpolitik im 21. Jahrhundert. Wien.
Ilgmann, Gottfried, 2001: Entherrschung und Chinese Walls im DB-Konzern: Eine Stellungnahme zu den Überlegungen der TaskForce zur Trennung von Netz und Transport, September 21. Hamburg.
IRG, 2005: FTA- Report on fixed termination access, IRG(05) 25. Brussels.
Jäger, Bernd, 1994: Postreform I und II: Die gradualistische Telekommunikationspolitik in Deutschland im Lichte der Positiven Theorie staatlicher Regulierung und Deregulierung. Köln.

Jahn, Eric/ Prüfer, Jens, 2004: Transit versus (Paid) Peering: Interconnection and Competition in the Internet Backbone Market. Paper presented at WIP Conference 2004, Berlin.
Jamison, Mark A., 2005: Leadership and the Independent Regulator, World Bank Policy Research Working Paper 3620, June. Gainesville.
Janßen, Bettina, 2001: Der Markt beginnt sich zu öffnen. In: Wirtschaftspsychologie, No. 2, pp. 36-37.
Kahn, Alfred E., 1988: The Economics of Regulation - Principles and Institutions. Cambridge, Mass.
Kahneman, Daniel, 2000: Evaluation by Moments: Past and Future. In: Choices, Values, and Frames, Kahneman, Daniel/ Tversky, Amos (eds.), Cambridge, Mass, pp. 693-708.
Kahneman, Daniel/ Knetsch, Jack L./ Thaler, Richard H., 1986: Fairness and the Assumptions of Economics. In: The Journal of Business, Vol. 59, No. 4, Part 2: The Behavioral Foundations of Economic Theory (October), pp. 285-300.
Kahneman, Daniel/ Tversky, Amos, 1991: Conflict Resolution: A Cognitive Perspective. In: Barriers to conflict resolution, Kenneth J. Arrow et al. (eds.), New York, pp. 44-62.
Kals, Elisabeth, 2003: Konfliktlösung und Mediation 1996-2003, Bibliographien zu Psychologie, No. 126. Zentrum für Psychologische Information und Dokumentation, Universität Trier (ed.). Norderstedt.
Kals, Elisabeth/ Kärcher, Juliane, 2001: Mythen in der Wirtschaftsmediation. In: Wirtschaftspsychologie, No. 2, pp. 17-27.
Kals, Elisabeth/ Webers, Thomas, 2001: Wirtschaftsmediation als alternative Konfliktlösung. In: Wirtschaftspsychologie, No. 2, pp. 10-16.
Kastenholz, Hans/ Benighaus, Christina, 2003: Information und Dialog bei der Standortsuche von Mobilfunkanlagen. Sozialministerium Baden-Württemberg (ed.). Stuttgart.
Kemp, Ray V./ Greulich, Tamsin, 2004: Working with the Community - Handbook on mobile telecoms community consultation for best siting practice. mobile operators association (moa) (ed.). London.
Keuter, Christof/ Waitschies, Markus, 2004: Der aktuelle Begriff: Die AEG-Novelle - mehr Wettbewerb auf der Schiene. Wissenschaftliche Dienste des Deutschen Bundestages, No. 43. Berlin.
Kiessling, T/ Blondeel, Y, 1998: The EU regulatory framework in telecommunications: A critical analysis. In: Telecommunications Policy, Vol. 22, pp. 571-592.
Kirchner, Christian, 2002: Offene Regulierungsprobleme bei Betreibervorauswahl und -vorauswahl (carrier selection) im Ortsnetz. Berlin.
Kirchner, Christian, 2003: Liberalisierungsindex Bahn 2002. Vergleich der Marktöffnung in den Eisenbahnmärkten der 15 Mitgliedsstaaten der Europäischen Union, der Schweiz und Norwegen. Berlin.
Kirchner, Christian, 2004: Liberalisierungsindex Bahn 2004. Vergleich der Marktöffnung der Eisenbahnmärkte der Mitgliedsstaaten der Europäischen Union, der Schweiz und Norwegen. Berlin.

Klatt, Sigurd, 1994: Stichwort Verkehrsinfrastruktur. In: Handwörterbuch der Raumforschung und Raumordnung. Hannover.

Knieps, Günter, 2004: Privatisation of Network Industries in Germany: A Disaggregated Approach, CESIfo Working Paper No. 1188, May. München.

Knieps, Günter/ Brunekreeft, Gert, 2000: Zwischen Regulierung und Wettbewerb - Netzsektoren in Deutschland. Heidelberg.

Koch, Michael, 2004: Powerline: Status und Perspektiven. Paper presented at TK-Tag Hessen, July 1. Wiesbaden.

Koenig, Christian/ Vogelsang, Ingo/ Winkler, Kay E., 2005: Marktregulierung im Bereich der Mobilfunkterminierung. Bonn.

Kohlstedt, Alexander, 2003a: Selbstregulierung im deutschen TK-Markt: Das Beispiel des AKNN. In: WIK-Newsletter, No. 52, pp. 3-5.

Kohlstedt, Alexander, 2003b: The Changing European Framework in Telecommunication. In: WIK-Newsletter, No. 52, pp. 17-20.

König, Ursula/ Prader, Thomas/ Zillessen, Horst, 2005: Abschlussdokumente Mediationsverfahren Flughafen Wien. Wien.

Korobkin, Russell/ Moffitt, Michael/ Welsh, Nancy, 2004: The Law of Bargaining, Research Paper No. 04-3. Los Angeles.

Köster, Dieter, 1999: Wettbewerb in Netzproduktmärkten. Wiesbaden.

Krummheuer, Eberhard, 2005a: Private Konkurrenten siegen über die Bahn: Energietochter der Deutschen Bahn verliert Prozess. In: Handelsblatt, January 28, p. 17.

Krummheuer, Eberhard, 2005b: Nachgefragt: Stéphane Richard. In: Handelsblatt, February 16, p. 17.

Kruse, Jörn, 1992: Mobilfunk in Deutschland: Ordnungspolitik und Marktstrukturen, WIK-Diskussionsbeitrag No. 94. Bad Honnef.

Külp, Bernhard, 1965: Theorie der Drohung. Köln.

Kurth, Matthias, 2002: Mobilfunk-Festnetz, Partnerschaft oder angespannte Konkurrenz? Paper presented at Forumsveranstaltung der RegTP, October 22. Bonn.

Kurth, Matthias, 2005: Voice over IP - Revolution oder Evolution auf dem TK-Markt? In: Multimedia und Recht, No. 3, supplement, pp. 3-7.

Kuthan, Jiri, 2004: VoIP/SIP in Mass Market 2003/2004, Paper presented at the DIHK, May 13. Berlin.

Langer, Ellen J., 1982: The illusion of control. In: Judgement under uncertainty: Heuristics and biases, Kahneman, Daniel/ Slovic, Paul/ Tversky, Amos (eds.), Cambridge, pp. 231-238.

Lapuerta, Carlos/ Tye, William B., 1999: Promoting effective competition through interconnection policy. In: Telecommunications Policy, Vol. 23, pp. 129-145.

Latzer, Michael et al., 2003: Regulation Remixed: Institutional Change through Self and Co-Regulation in the Mediamatics Sector. In: Communications & Strategies, No. 50, pp. 127-157.

Lavender, Tony, 2005: BT to form separate access division. In: Ovum Straight Talk, February 4, London.

Lavie, Dovev, 2003: The competitive advantage of interconnected firms: an extension of the resource-based view. Austin.

Leib, Volker, 2002: ICANN und der Konflikt um die Internet-Ressourcen: Institutionenbildung im Problemfeld Internet Governance zwischen multinationaler Staatstätigkeit und globaler Selbstregulierung. Konstanz.

Lewicki, Roy J. et al., 1994: Negotiation. Boston.

Leykauf, F., 2000: Wirtschaftsmediation als Verfahren aussergerichtlicher Konfliktregelung: Anforderungen, Umsetzung, Verbreitung. Diplomarbeit. Wiesbaden.

Ludewig, Johannes, 2005: Europa braucht leistungsfähige Schienenkorridore. In: Frankfurter Allgemeine Zeitung, March 29, p. 19.

Lunden, Ingrid, 2005: Disruptive or complementary? In: Total Telecom Magazine, May, p. 36.

Lyon, Thomas P./ Huang, Haizhou, 2002: Legal Remedies for Breach of the Regulatory "Contract". In: Journal of Regulatory Economics, Vol. 22, No. 2 (September), pp. 107-132.

Machatschke, Michael, 2002: Attacke auf die Bahn. In: manager magazin, No. 4, pp. 156-165.

Mähler, Hans-Georg/ Mähler, Gisela, 2000: Gerechtigkeit in der Mediation. In: Gerechtigkeit im Konfliktmanagement und in der Mediation, Dieter, Anne/ Montada, Leo/ Schulze, Annedore (eds.), Frankfurt, pp. 9-36.

Mai, Christine, 2004: Lange Prozesse lähmen Telekomanbieter. In: Financial Times Deutschland, July 6, p. 4.

Mandewirth, Sven-Oliver, 1997: Transaktionskosten von Handelskooperationen: Ein Effizienzkriterium für Verbundgruppen und Franchise Systeme. Heidelberg.

Manzini, Paola/ Mariotti, Marco, 2002: Arbitration and Mediation: An Economic Perspective, IZA Discussion Paper No. 528, July. Bonn.

Maoz, Zeev, 1989: Power, Capabilities, and Paradoxical Conflict Outcomes. In: World Politics, Vol. 41, No. 2 (January), pp. 239-266.

March, James G., 1963: The Power of Power. In: Varieties of Political Theory, Easton, David (ed.), Englewood Cliffs, NJ, pp. 39-70.

Märker, Oliver/ Trénel, Matthias, 2003: Online-Mediation: Neue Medien in der Konfliktvermittlung - mit Beispielen aus der Politik und Wirtschaft. Berlin.

Marsden, Christopher, 2005: Co-Regulation in European Media and Internet Sectors. In: Multimedia und Recht, No. 3.

Matthies, Wolfgang/ Schinkel, 1991: Die Erneuerung der Telekommunikationsnetze der Bundesbahn und der Reichsbahn auf Basis der ISDN-Konzeption. In: Signal + Draht, Vol. 83, No. 6, pp. 141-147.

Mc Nutt, Patrick A., 1996: The economics of public choice. Cheltenham.

McKinsey, 2002: Regulierungsmanagement Deutsche Bahn - Diskussion mit Herrn Dr. Hedderich. Berlin.

Mechnig, Stefan Paul, 2004: Telekom und Wettbewerber gehen in Rechungsstreit aufeinander zu. In: vwd, February 5.

Mehr Bahnen, 2003: Scheitert der Wettbewerb auf der Schiene? Berlin.

Mehr Bahnen, 2005: Schwarzbuch Diskriminierung. Berlin.

Michalowitz, Irina, 2004: EU Lobbying - Principals, Agents and Targets: Strategic interest intermediation in EU policy-making, Public Affairs und Politikmanagement, No. 4. Potsdam.
Miller, James C. III, 1985: The FTC and Voluntary Standards: Maximizing the net benefits of self-regulation. In: Cato Journal, Vol. 4, No. 3 (Winter), pp. 897-903.
Miller, Robert C.B., 2003: railway.com - Parallels between the early British railways and the ICT revolution. London.
Mobile Operators Association, 2004: Base Station and Masts. London.
Monopolkommission, 2004a: Telekommunikation und Post 2003: Wettbewerbsintensivierung in der Telekommunikation - Zementierung des Postmonopols Sondergutachten der Monopolkommission gemäß §81 Abs. 3 Telekommunikationsgesetz und § 44 Postgesetz. Bonn.
Monopolkommission, 2004b: Zur Reform des Telekommunikationsgesetzes - Sondergutachten der Monopolkommission gemäß §44 Abs. 1 Satz 4 GWB. Bonn.
Montada, Leo, 2000: Gerechtigkeit und Rechtsgefühl in der Mediation. In: Gerechtigkeit im Konfliktmanagement und in der Mediation, Dieter, Anne/ Montada, Leo/ Schulze, Annedore (eds.), Frankfurt, pp. 37-62.
Mueller, Dennis C., 2003: Public Choice III. Cambridge.
Nalebuff, Barry/ Brandenburger, Adam, 1996: Coopetition - kooperativ konkurrieren. Frankfurt.
Nash, John F., 1950: The Bargaining Problem. In: Econometrica, Vol. 18, No. 2 (April), pp. 155-162.
Nett, Lorenz/ Neumann, Karl-Heinz/ Vogelsang, Ingo, 2004: Geschäftsmodelle und konsistente Entgeltregulierung. Studie für die Regulierungsbehörde für Telekommunikation und Post. April 2. Bad Honnef.
NZZ (Neue Züricher Zeitung), 2004: Mehr Staatskontrolle für die britische Bahn. In: Neue Züricher Zeitung, July 16.
Neuhoff, Lothar, 2003: Interview, October 27 in Berlin. FDP-Bundestagsfraktion, Verkehrsreferent.
Neuhoff, Lothar, 2005: Interview, January 3 (telephone-interview). FDP-Bundestagsfraktion, Verkehrsreferent.
Neumann, Karl-Heinz, 2002: Konsistentes Entgeltregulierungsregime im Zusammenhang mit der Betreiber(vor)auswahl im Ortsnetz. Bad Honnef.
Neumann, Karl-Heinz, 2004: Konsistente Entgeltregulierung. In: WIK-Newsletter, No. 54, Bad Honnef, pp. 1-2.
Neumann, Karl-Heinz/ Stamm, Peter, 2002: Wettbewerb in der Telekommunikation - Wie geht es weiter? Bad Honnef.
Neuscheler, Tillmann, 2005: Ein Regulierer für alle Netze. In: Frankfurter Allgemeine Zeitung, March 27, p. 45.
Newbery, David M., 2000: Privatization, Restructuring, and Regulation of Network Utilities. Cambridge.
Niemeier, Stefan, 2002: Die deutsche UMTS-Auktion. Wiesbaden
Nolte, Norbert/ König, Annegret, 2005: Konsistente Entgeltregulierung im neuen TKG. In: Multimedia und Recht, No. 8, pp. 512-517.

North, Douglass C., 1992: Institutionen, institutioneller Wandel und Wirtschaftsleistung. Freiburg.
O'Connor, Kathleen/ Arnold, Josh, 2002: The Shadow of the Past: How Past Negotiation Performance Affects Future Negotiation Performance. Cornell.
OECD, 1998: Railways: Structure, Regulation and competition policy, DAFFE/CLP(98)1. Paris.
Oftel, 2000: Resolving disputes between Fixed Telecommunications Service Operators. London.
Oftel, 2003: Reviews of dispute procedure schemes: Draft guidelines Draft guidelines issued by the Director General of Telecommunications. London.
Otelo, 2003: Introducing: The Telecommunications Ombudsman Service. The Ombudsman's First Report. London.
Ott, Klaus, 2004: Connex beendet Attacken auf die Deutsche Bahn. In: Süddeutsche Zeitung, November 19.
Ovum, 2002: GSM Europe Code of Conduct for Information on International Roaming Retail Prices. Code of Conduct Monitoring: Results for first year of implementation (December 2001 - October 2002). London.
Parsons, Talcott, 1966: The Political Aspect of Social Structure and Process. In: Varieties of Political Theory, Easton, David (ed.), Englewood Cliffs, N.J., pp. 71-112.
Pascale, Richard T. et al., 2002: Chaos ist die Regel - Wie Unternehmen Naturgesetze erfolgreich anwenden. München.
Pelkmans, Jacques, 2001: Making the EU network markets competitive. In: Oxford Review of Economic Policy, Vol. 17, No. 3, pp. 432-456.
Peltzman, Sam/ Winston, Clifford, 2000: Deregulation of Network Industries: What's next? Washington, D.C.
Pen, J., 1952: A General Theory of Bargaining. In: American Economic Review, Vol. 42, No. 1 (March), pp. 24-42.
Pergande, Frank, 2004: Begegnungen: Der junge Bahnmanager. In: Frankfurter Allgemeine Zeitung, March 2, p. 18.
Perrucci, Antonio/ Cimatoribus, Michela, 1997: Competition, convergence and asymmetry in telecommunications regulation. In: Telecommunications Policy, Vol. 21, pp. 493-512.
Pesch, Stephan, 2003: Regulierung: Was kommt auf die Energieunternehmen zu? In: ew, Vol. 102, No. 22, pp. 30-33.
Petersdorff, Wienand von, 2005: "Das Internet ist die Zukunft des Telefons". In: Frankfurter Allgemeine Zeitung, January 16, p. 33.
Piepenbrock, Hermann-Josef, 2004: Antrag der 01051 Telecom GmbH auf Anordnung der Zusammenschaltung öffentlicher Telekommunikationsnetze mit der Vodafone D2 GmbH. Düsseldorf.
Pies, Ingo, 2000: Theoretische Grundlagen demokratischer Wirtschafts- und Gesellschaftspolitik - Der Beitrag von Ronald Coase. In: Ronald Coase' Transaktionskostenansatz, Pies, Ingo/ Leschke, Martin (eds.), Tübingen, pp. 1-29.
Poel, Martijn, 2002: Models of self and co-regulation, Paper presented at Malta Regional Forum October 28 - 29. Malta.

Poitras, Jean, 2005: A study on the emergence of cooperation in mediation. In: Negotiation Journal, April, pp. 281-300.
Poitras, Jean/ Bowen, Robert E., 2002: A Framework for Understanding Consensus-Building Initiation. In: Negotiation Journal, July, pp. 211-232.
Poitras, Jean/ Bowen, Robert E./ Byrne, Sean, 2003: Bringing Horses to Water? Overcoming Bad Relationships in the Pre-Negotiating Stage of Consensus Building. In: Negotiation Journal, July, pp. 251-263.
Polatschek, Klemens, 2004: Es fährt ein Zug nach Irgendwo. In: Frankfurter Allgemeine Zeitung, March 28, pp. 70-71.
PriceWaterhouseCoopers, 2005: Commercial Dispute Resolution – Konfliktbearbeitungsverfahren im Vergleich. Frankfurt.
Projektgruppe Gerichtsnahe Mediation in Niedersachsen, 2004: Erfahrungen aus der Projektpraxis. Hannover.
PSPC, 2003: Drittes Gesetz zur Änderung Eisenbahnrechtlicher Vorschriften (3. AEG-Novelle), Gutachten zur Bewertung des Referentenentwurfs. Berlin.
PSPC, 2004: Wettbewerb im Schienenverkehr - Kaum gewonnen, schon zerronnen? Gutachten im Auftrag von MehrBahnen. Berlin.
Putz & Partner, 2004: Telekommunikationsverbände in Deutschland. Hamburg.
Racine, Jérôme/ Auer, Andre, 2001: Wie verhandelt man multilateral erfolgreich? In: Neue Züricher Zeitung, September 22.
Racine, Jérôme/ Winkler, Klaus, 2002: Konfliktlösungsansätze im TK-Markt - Hilft das ADR-Grünbuch der EU weiter? In: Multimedia und Recht, No. 12, pp. 794-798.
Raiffa, Howard, 1982: The Art and Science of Negotiation. Cambridge.
Rawls, John, 1971: A theory of justice. Cambridge.
RegTP (Regulierungsbehörde für Telekommunikation und Post), 2002: Annual Report 2002. Bonn.
RegTP (Regulierungsbehörde für Telekommunikation und Post), 2003a: Terminierungsentgelte alternativer Teilnehmernetzbetreiber im Festnetz: Stellungnahmen. Bonn.
RegTP (Regulierungsbehörde für Telekommunikation und Post), 2003b: Annual Report 2003. Bonn.
RegTP (Regulierungsbehörde für Telekommunikation und Post), 2004a: VoIP: Stellungnahmen. Bonn.
RegTP (Regulierungsbehörde für Telekommunikation und Post), 2004b: Verbindungsleistungen im Festnetz. In: Official Journal, No. 18, Bonn.
RegTP (Regulierungsbehörde für Telekommunikation und Post), 2005a: Annual Report 2004. Bonn.
RegTP (Regulierungsbehörde für Telekommunikation und Post), 2005b: Bundesnetzagentur senkt Entgelte für Teilnehmerdaten, press-release, August 17. Bonn.
Rehbein, Jochen, 1995: International sales talk. In: The Discourse of Business Negotiation, Ehlich, Konrad/ Wagner, Johannes (eds.), Berlin, pp. 67-102.
Rickert, Beate, 2004: Das Breitbandkabel als Wettbewerbsinfrastruktur im Bereich Internet?, paper presented at DIHK, May 13. Berlin.

Rieck, Olaf, 2005: The impact of service innovation on corporate performance: an investigation into the mobile telecommunications service industry. Singapore.

Riedel, Donata, 2002: Telekom will konstruktiver mit Regulierer verhandeln. In: Handelsblatt, November 13.

Ritter, Eva-Maria, 2004: Deutsche Telekommunikationspolitik 1989-2003 - Aufbruch zu mehr Wettbewerb. Ein Beispiel für wirtschaftliche Strukturreformen, Beiträge zur Geschichte des Parlamentarismus und der politischen Parteien. Düsseldorf.

Rochlitz, Karl-Heinz, 2002: Verpasster Konsens beim Ausschreibungswettbewerb. In: Nahverkehrspraxis, No. 10, p. 1.

Rochlitz, Karl-Heinz, 2003: Interview, April 3 in Frankfurt. Connex Verkehr GmbH, Marketing und Kooperationen.

Rochlitz, Karl-Heinz, 2005: DB & Connex, Dossier. Mettmann.

Rochlitz, Karl-Heinz/ Winkler, Klaus, 2005: RegTP+EBA=Super-Regulierer? - RegTP übernimmt die Aufsicht auch über die Schiene. In: Multimedia und Recht, No. 7, pp. XI-XIII.

Rodini, Mark/ Ward, Michael R./ Woroch, Glenn A., 2003: Going mobile: substitutability between fixed and mobile access. In: Telecommunications Policy, Vol. 27, pp. 457-476.

Rothschild, K.W., 1971: Power in Economics. Harmondsworth.

Rubinstein, Ariel, 1982: Perfect Equilibrium in a Bargaining Model. In: Econometrica, Vol. 50, No. 1 (January), pp. 97-109.

Ruhle, Ernst-Olav, 2005: RegTP - Allein zuhaus? In: Piepenbrock und Schuster, Newsletter, No. 24, Düsseldorf, p. 11.

Ruhle, Ernst-Olav/ Lichtenberger, Ewald/ Kittl, Jörg, 2005: Erste Erfahrungen mit dem TKG in Österreich - Eine Analyse der Gesetzesanwendung durch die Regulierungsbehörde. In: Medien und Recht, No. 1, Wien, pp. 1-18.

Schaaffkamp, 2004: Paper presented at DVWG in Frankfurt, April 7. Frankfurt.

Schäfer, Ralf G., 2003: Telekommunikationswettbewerb in Deutschland aus der Nachfragerperspektive. In: WIK-Newsletter, No. 52, pp. 6-9.

Scheerer, M./ Nonnast, Th., 2005: Brüssel greift deutschen Regulierer an. In: Handelsblatt, February 11, p. 3.

Schelling, Thomas C., 1956: An Essay on Bargaining. In: American Econonomic Review, Vol. 46, No. 3 (June), pp. 281-306.

Schelling, Thomas C., 1957: Bargaining, Communication, and Limited War. In: Conflict Resolution, Vol. 1, No. 1 (March), pp. 19-36.

Schelling, Thomas C., 1958: The Strategy of Conflict Prospectus for a Reorientation of Game Theory. In: The Journal of Conflict Resolution, Vol. 2, No. 3 (September), pp. 203-264.

Schelling, Thomas C., 1980: The Strategy of Conflict. Harvard.

Scherer, Joachim, 2002: Die Umgestaltung des europäischen und deutschen Telekommunikationsrechts durch das EU-Richtlinienpaket Teil I-III. In: Kommunikation & Recht, No. 6-8, pp. 273-289; 329-346; 385-398.

Scherer, Joachim/ Schimanek, Peter, 2002: Rechtsfragen elektromagnetischer Felder ("Elektrosmog"). In: Jahrbuch des Umwelt- und Technikrechts 2002, Reinhard Hendler et al. (eds.), Berlin, pp. 295-317.

Scheurle, Klaus-Dieter, 2003: Interview, February 2 in Frankfurt. Credit Suisse First Boston (Europe) Limited, Managing Director, Mitglied der Geschäftsleitung.

Schintler, Laurie et al., 2004: Complex network phenomena in telecommunication systems, Tinbergen Institure Discussion Paper, TI 2004-118/3. Tinbergen.

Schmidt, Frank, 2004: Grundsatzartikel zur aktuellen wettbewerbspolitischen Diskussion um Optionstarife und Paketangebote im Telekommunikationsmarkt. Bonn.

Schmidt, Holger, 2005a: Elf Wettbewerber gegen Titelverteidiger Telekom. In: Frankfurter Allgemeine Zeitung, March 10, p. 14.

Schmidt, Holger, 2005b: "Die Menschen sollen kostenlos telefonieren". In: Frankfurter Allgemeine Zeitung, April 11, p. 19.

Schmidt, Holger/ Winkelhage, Johannes, 2005: "Die Tricks der Stromnetzbetreiber sind uns bekannt". In: Frankfurter Allgemeine Zeitung, February 2, p. 13.

Schneider, Jürgen, 1995: Die Privatisierung der Deutschen Bundes- und Reichsbahn. Wiesbaden.

Scholl, Mechthild, 1998: Telekommunikation zwischen Regulierung und Deregulierung. Bonn.

Schönborn, Gregor/ Wiebusch, Dagmar, 2002: Public Affairs Agenda. Neuwied.

Schuppert, Gunnar Folke, 1990: Resümee und Ausblick. In: Konfliktbewältigung durch Verhandlungen Vol. 1 Informelle und mittlerunterstützte Verhandlungen in Verwaltungsverfahren, Hoffmann-Riem, Wolfgang/ Schmidt-Aßmann, Eberhard (eds.), Baden-Baden, pp. 327-335.

Schütz, Raimund, 2005a: Kommunikationsrecht: Regulierung von Telekommunikation und elektronischen Medien. München.

Schütz, Raimund, 2005b: DTAG-Regulierung: Neues Spiel - Neues Glück? In: Multimedia und Recht, No. 2, p. VI.

Schütz, Raimund, 2005c: Präsident der Bundesnetzagentur - ein Traumjob? In: Multimedia und Recht, No. 6, p. VIII.

Schweinsberg, Ralf, 2002: Diskriminierungsfreier Netzzugang auf der Eisenbahn aus Sicht der Aufsichtsbehörde Eisenbahn-Bundesamt, Presentation at the workshop of DVWG and IVE, October 17. Bonn.

Schweizer, Adrian, 2002: Kooperatives Verhandeln - die Alternative zum (Rechts-)Streit. In: Handbuch Mediation, Haft, Fritjof/ Schlieffen, Katharina Gräfin von (eds.), München, pp. 210-235.

Schwerhoff, Ulrich, 2002: Sachverständigengutachten zum Thema Carrier Selection im Ortsnetz. Wiesbaden.

Seeley, Elizabeth/ Gardner, Wendi/ Thompson, Leigh, 2002: Interdependent Self-Construals Increase the Benevolent Use of Power in a Dispute Resolution. Evanston.

Shapiro, Carl/ Varian, Hal R., 1999: Information Rules - A Strategic Guide to The Network Economy. Boston.

Sicker, Douglas C./ Lookabaugh, Tom, 2004: A model for emergency service of VoIP through certification and labeling, NET Institute, working paper No. 04-19. Boulder.

Siedenbiel, Christian, 2003: Connex stellt Ersatz für Interregio wieder ein. In: Frankfurter Allgemeine Zeitung, October 4, p. 78.
Simon, Herbert A., 1957: Models of Man - Social and Rational. New York.
Simon, Herbert A., 1966: Political Research: The Decision-Making Framework. In: Varieties of Political Theory, Easton, David (ed.), Englewood Cliffs, N.J., pp. 15-24.
Simon, Herbert A., 1982: Models of Bounded Rationality Vol. 2: Behavioral Economics and Business Organization. Cambridge, Mass.
Sisalem, Dorgham, 2004: VoIP: Why now?, paper presented at DIHK, May 13. Berlin.
SPD, 2004: Innovation und Gerechtigkeit - Telekommunikation als Innovationsmotor für Deutschland, Veranstaltung der SPD-Bundestagsfraktion am 4. November 2004 in Berlin. Berlin.
Spiller, Kristina, 2004: Tchibo heizt Mobilfunk-Konkurrenz an. In: Financial Times Deutschland, July 12.
Ståhl, Ingolf, 1972: Bargaining Theory. Stockholm.
Stefanadis, Christodoulos, 2003: Self-Regulation, Innovation, and the Financial Industry. In: Journal of Regulatory Economics, Vol. 23, No. 1 (January), pp. 5-25.
Stieber, Carolyn, 2000: 57 Varieties: Has the Ombudsman Concept Become Diluted? In: Negotiation Journal, January, pp. 49-57.
Stigler, George J., 1971: The Theory of Economic Regulation. In: The Bell Journal of Economics and Management Science, Vol. 2, No. 1 (Spring), pp. 3-21.
Stinnes, 2003: Stinnes: Geschäftsjahr 2002 geprägt duch die Übernahme der Bahn, Press release, March 24. Berlin.
Stoffels, Markus, 2004: Rechtsgutachten zu den Standardverträgen zwischen der DTAG und den alternativen Diensteanbietern. Köln.
Strauch, Manfred, 1993: Lobbying: Wirtschaft und Politik im Wechselspiel. Wiesbaden.
Stuhmcke, Anita, 2002: The rise of the Australian Telecommunications Industry Ombudsman. In: Telecommunications Policy, Vol. 26, pp. 69-85.
Stüwe, Heinz, 2004: Telekom macht Inkasso für Wettbewerb - Blocktarife werden über Telefonrechnung abgerechnet. In: Frankfurter Allgemeine Zeitung, March 10, p. 14.
Susskind, Lawrence/ McKearnan, Sarah/ Thomas-Larmer, Jennifer, 1999: The Consensus Building Handbook: A Comprehensive Guide to Reaching Agreement. Thousand Oaks.
Tartler, Jens, 2005: Niederlage für Mehdorns Kritiker. In: Financial Times Deutschland, August 25.
Tele2, 2005: The Monopoly Challenger, March. Stockholm.
Telekom-Control-Kommission, 2005: Ergebnisse der Konsultation: Ermittlung der Kosten der effizienten Leistungsbereitstellung für Terminierung in Mobilfunknetzen. Wien.

Tirole, Jean, 2004: Telecommunications and competition. In: The economics of antitrust and regulation in telecommunications: perspectives for the new European Regulatory Framework, Buigues, Pierre A./ Rey, Patrick (eds.), Cheltenham, pp. 260-265.
TK-dialog, 2004: Workshop "Das neue TKG - Umsetzung in die Praxis", July 21. Frankfurt.
Tomik, Stefan, 2002: Die Reform der Bahnreform. Berlin.
Tversky, Amos/ Kahneman, Daniel, 1986: Rational Choice and the Framing of Decisions. In: The Journal of Business, Vol. 59, No. 4 (October), Part 2: The Behavioral Foundations of Economic Theory, pp. 251-278.
Ury, William L., 1992: Schwierige Verhandlungen. München.
Vanberg, Margit A., 2002: Competition in the German Broadband Access Market. In: Discussion Paper No. 02-80, Centre for European Economic Research (ZEW). Mannheim.
VATM, 2001: Aktuelle Markteintrittsbarrieren auf dem deutschen Telekommunikationsmarkt. Köln.
VATM, 2002: Stellungnahme zum 8. Implementierungsbericht. Köln.
VATM, 2004: Medienresonanzanalyse. Köln.
VATM, 2005: Deutsche Telekom und VATM einigen sich auf neue Abrechnungsmodalitäten, press-release August 2. Köln.
Vaubel, Roland, 1991: The political economy of international organizations: a public choice approach. Boulder.
VCD, 2004a: Zehn Jahre Bahnreform - VCD-Position. Bonn.
VCD, 2004b: VCD Bahntest 2004, Hintergrundinformationen und Ergebnisse, April 2004. Bonn.
Verbraucher-Zentrale NRW, 2004: Mehr Rechte für Fahrgäste - Schlichtungsstelle Nahverkehr: Projektbericht 2003. Düsseldorf.
Vogelsang, Ingo, 2001: Neue Probleme der Preisregulierung von Endkundenleistungen - Preisbündelung auf Telekommunikationsmärkten, paper presented at WIK Workshop, December 14. Bonn.
Vogelsang, Ingo, 2002: Die Zukunft der Entgeltregulierung im deutschen Telekommunikationssektor. München.
Vogelsang, Ingo/ Mitchell, Bridger M., 1997: Telecommunications Competition: The Last Ten Miles. Cambridge.
Voigt, Stefan, 2002: Institutionenökonomik. München.
Waesche, Niko Marcel, 2003: Internet entrepreneurship in Europe: venture failure and the timing of telecommunications reform. Cheltenham.
Wagner, Johannes, 1995: What makes a discourse a negotiation? In: The Discourse of Business Negotiation, Ehlich, Konrad/ Wagner, Johannes (eds.), Berlin, pp. 8-36.
Wall, James A. / Lynn, Ann, 1993: Mediation: A Current Review. In: The Journal of Conflict Resolution, Vol. 37, No. 1 (March), pp. 160-194.
Wall, James A. / Stark, John B./ Standifer, Rhetta L., 2001: Mediation: A Current Review and Theory Development. In: The Journal of Conflict Resolution, Vol. 45, No. 3 (June), pp. 370-391.

Warren, J. Samuels, 1979: The Economy as a System of Power. Warren, J. Samuels (ed.), New Jersey.
Wegmann, Winfried, 2001: Regulierte Marktöffnung in der Telekommunikation: Die Steuerungsinstrumente des Telekommunikationsgesetztes (TKG) im Lichte "regulierter Selbstregulierung", Wirtschaftsrecht der internationalen Telekommunikation, No. 45. Baden-Baden.
Wendel, Thomas H., 2004: DTAG: Buntes Durcheinander. In: Berliner Zeitung, February 11.
Wentzel, Dirk, 2002: Principles of Self-Regulation, Volkswirtschaftliche Beiträge, No. 1. Marburg.
Werner, Jan, 2002: Interview, November 6 in Berlin. Kompetenzzentrum Wettbewerb im HVV.
Werner, Jan et al., 2004: Zur Anwendbarkeit der VO (EWG) No. 1191/69 in Deutschland - Gutachten im Auftrag des RMV. Berlin.
Wiesmann, Bettina M./ Steinbach, Anke, 2002: Wettbewerb im ÖPNV - Herausforderungen und Chancen für die Kommunen. In: Nahverkehrspraxis, No. 9 and 10, pp. 44-46 and 9-11.
WIK, 2005: Zur Konsolidierungsdiskussion im deutschen Postmarkt: Möglichkeiten und Entgelte für Netzzugang. Bericht für den Bundesverband Internationaler Express- und Kurierdienste (BIEK). Bad Honnef.
Williamson, Oliver E., 1993: Transaktionskostenökonomik, Ökonomische Theorie der Institutionen, Vol. 3. Hamburg.
Winkelhage, Johannes, 2005: Wer mit der Internettelefonie sein Geld verdient. In: Frankfurter Allgemeine Zeitung, May 9, p. 20.
Winkler, Klaus, 2003: Strafe für unbequeme Kunden - Der "Selbstzahlerzuschlag" der DTAG. In: Multimedia und Recht, No. 3, pp. XXIII-XXV.
Winkler, Klaus, 2005: Mediation im TKG - Eine ökonomische Betrachtung. In: Multimedia und Recht, No. 3, pp. X-XII.
Winterstetter, Bernhard, 2002: Ökonomische Aspekte der Mediation. In: Handbuch Mediation, Haft, Fritjof/ Schlieffen, Katharina Gräfin von (eds.), München, pp. 510-526.
Wirtschaftsrat, 2003: Minutes of the meeting on September 24. Berlin.
Wirtschaftsrat, 2004: Minutes of the meeting on March 25. Berlin.
Wissenschaftlicher Beirat beim Bundesminister für Verkehr, Bau- und Wohnungswesen, 2002: Trennung von Netz und Transport im Eisenbahnwesen. In: Internationales Verkehrswesen, Vol. 54, No. 6, pp. 260-264.
Wittschier, Bernd M., 2001: So wirkt Mediation. In: Wirtschaft & Weiterbildung, June.
Wünschmann, Christoph, 2005: Interview, January 20 in Berlin. Hogan & Hartson Raue LLP.
Yan, Xu, 2001: The impact of the regulatory framework on fixed-mobile interconnection-settlements: the case of China and Hong Kong. In: Telecommunications Policy, Vol. 25, pp. 515-532.
Yankelovich, Daniel, 1982: Lying well is the best revenge. In: Psychology Today, Vol. 16, pp. 5-6.

Young, David, 2002: The meaning and role of power in economic theories. In: A Modern Reader in Institutional and Evolutionary Economics: Key Concepts, Geoffry M. Hodgson (ed.), Cheltenham, pp. 48-61.

Young, Mark A., 2001: Rational Games - A Philosophy of Business Negotiation from Practical Reason. Westport, CT.

Ypsilanti, Dimitri/ Xavier, Patrix, 1998: Towards next generation regulation. In: Telecommunications Policy, Vol. 22, pp. 643-659.

Zartman, I. William, 1974: The Political Analysis of Negotiation: How Who Gets What and When. In: World Politics, Vol. 26, No. 3 (April), pp. 385-399.

Zartman, I. William, 1978: The Negotiation Process. Beverly Hills.

Zartman, I. William/ Rubin, Jeffrey Z., 2000: Power and negotiation. Michigan.

Zuberbühler, Christa, 2004: WirtschaftsMediation - Durch Konsens zum Erfolg. Zürich.